# Cotton Cultivation and Child Labor in Post-Soviet Uzbekistan

# Cotton Cultivation and Child Labor in Post-Soviet Uzbekistan

Bilal Bhat

LEXINGTON BOOKS
Lanham • Boulder • New York • London

Published by Lexington Books
An imprint of The Rowman & Littlefield Publishing Group, Inc.
4501 Forbes Boulevard, Suite 200, Lanham, Maryland 20706
www.rowman.com

Unit A, Whitacre Mews, 26-34 Stannary Street, London SE11 4AB



British Library Cataloguing in Publication Information Available

**Library of Congress Cataloging-in-Publication Data**

Bhat, Bilal.
Cotton cultivation and child labor in post-Soviet Uzbekistan / Bilal Bhat.
pages cm
Includes bibliographical references and index.
ISBN 978-0-7391-9478-2 (cloth : alk. paper) -- ISBN 978-0-7391-9479-9 (electronic)
1. Cotton growing--Uzbekistan. 2. Cotton farmers--Uzbekistan. 3. Child labor--Uzbekistan. I. Title.
HD9086.U92B43 2015
331.3'1--dc23

2015019137

∞ ™ The paper used in this publication meets the minimum requirements of American National Standard for Information Sciences Permanence of Paper for Printed Library Materials, ANSI/NISO Z39.48-1992.

Printed in the United States of America

# Table of Contents

# Acknowledgments

I have accumulated a large number of debts in preparing this work. Without my teacher Mr. Tareak Ahmad Rather's sustained attention, searching discussion, and painstaking comment on various themes, the study would never have become what it is. B. A. Dabla, G. R. Jan, G. M. Mir, M. A. Mir, G. N. Khaki, and R. Makhdumi also deserve special thanks for their timely assistance. I thank M. A. Kaw and M. A. Wani of the University of Kashmir for taking personal pains while assisting and supporting me for administrative and academic purposes. I am thankful to the Area Study Programme of the Centre of Central Asian Studies, University of Kashmir, India, for financing my field study in Uzbekistan. From Uzbekistan I express heartfelt and most sincere gratitude to Vallery Khan, Azad Shamatov, and Ismat Bekmuratov. I must also mention the contribution of my friends, particularly Inam, Aijaz, Shahzad, Humaira, Tanveer, Shahnawaz, Sajad, Nilofar, and Owais through their spontaneous generosity and empathy.

I appreciate all the help and cooperation I got from faculty members and students of the various national and state institutes of Uzbekistan. I would like to express my heartfelt gratitude to my Uzbek friends Helga Frey, Malika Islamoya, Sator, Naeila, Alishree, Shahzod, Bahzod, Nilufar, Nargiza, Dilafroz, Ali, Islam, Dilshad, Barno, and many others for providing a stimulating and supportive environment for my research and writing. My sincere thanks to child laborers in the cotton industry, human rights activists, senior citizens, school teachers, and others who extended their cooperation in furnishing the information needed for the present study. I am beholden to Sabah Gulamali and Mr. Brian Hill of Lexington Books for reviewing my manuscript and giving me very helpful insights for improving the quality of the manuscript. This venture would not have been possible but for love, sacrifice, patience, and encouragement of my family—Ali, Zoon, Gousiya, Huriah, and Hareem in all my academic pursuits. I have no words to express my thanks to them. Lastly, I record my sincere and grateful thanks to all those authors and writers from whose writings and works I have benefitted.

# Abbreviations

| | |
|---|---|
| AFDE | Association of Farm and Dekhkan Economies |
| BBC | British Broadcasting Service |
| CCAS | Centre of Central Asian Studies |
| CEE | Central and Eastern Europe |
| CIS | Commonwealth of Independent States |
| CM | Cabinet of Ministers of the Republic of Uzbekistan |
| CRC | Convention on the Rights of the Child |
| CRU | Constitution of the Republic of Uzbekistan |
| EJF | Environmental Justice Foundation |
| EU | European Union |
| FSU | Former Soviet Union |
| GDP | gross domestic product |
| GRP | gross regional product |
| GSP | generalized system of preferences |
| HDI | human development index |
| HRSU | Human Rights Society of Uzbekistan |
| HRWCA | Human Rights Watch in Central Asia |
| ICAC | International Cotton Advisory Committee |
| IFI | international financial institution |
| ILO | International Labor Organization |
| ILRF | International Labor Rights Forum |
| IWPR | Institute of War and Peace Reporting |
| MDG | millennium development goals |
| MICS | multiple-indicator cluster survey |
| MLSP | Ministry of Labor and Social Protection of the Republic of Uzbekistan |
| MTPs | machine tractor parks |
| NAP | national agricultural production |
| NGO | nongovernment organization |

| | |
|---|---|
| RUz | Republic of Uzbekistan |
| TEA | temporary employment agency |
| TFR | total fertility rate |
| UN | United Nations |
| UNDP | United Nations Development Program |
| UNESCO | United Nations Educational, Social and Cultural Organization |
| UNICEF | United Nations Children's Emergency Fund |
| USSR | Union of Soviet Socialist Russia |
| UTC | urban-type community |
| UzSSR | Uzbek Soviet Socialist Republic |

---

Association of Farm and Dekhkan Economies Labor Organization Labor Rights Forum Ministry of Labor and Social Protection of the Republic of Uzbekistan Uzbek Soviet Socialist Republic

# ONE

# Introduction

The pervasiveness of child labor is not a recent phenomenon. In ancient times, children were put to arduous labor in houses and in the fields. Performances of tasks such as tending of cattle, collection of grass and fuel, and so on by children freed the adult members of the family for more productive work. There was no social taboo against children working along with their parents. It was this factor of child labor that strongly established family and kinship ties in occupations. '

In poor urban areas today, many parents cannot make any investment in their children's development and are also reluctant to support them. They want their children to find work for themselves as early as possible. In developing countries, children have always been the responsibility of the primary institution of society, viz. the family. But the traditional care lavished on children, has suffered erosion because of new constraints, and consequently abuse and exploitation of children has become common place. The plight of the child has been further aggravated by endemic and entrenched poverty. In spite of constitutional measures and a number of welfare-development schemes for them, a sizeable proportion of the youth population is still working as wage laborers/apprentices, paid or unpaid, in both urban and rural areas in very deplorable and inhuman conditions.

The problems of child labor became pronounced with the advent of industrial era. In the early phase of industrialization in the West, child labor suffered brutal and unprecedented exploitation. Being in the labor force, such children were denied educational opportunities and their physical, mental, and intellectual development was hampered. The working children of today suffer the same types of privations. They generally remain unskilled, underpaid, and underprivileged throughout life; their physical and social mobility is restricted and they fall into a vicious and cumulative cycle of poverty, ill-health, underemployment, and unemployment.[1] The multidimensional implications of child labor for children, parents, and families have been recognized as a social evil.[2]

Defining child labor, however, is not as simple and straightforward as it may appear because it encompasses three difficult-to-define concepts: "child," "work," and "labor." "Work is the application of human energy to things; which application converts, maintains, or adds value to the workers, the thing worked on, and the system in which the work is performed." However, work is not just a mechanical function of energy expenditure. It is energy directed to explicit goals. In its holistic approach work be taken as a human activity applying human energy and directed to some goal. Work is always performed as wage employment or for some kind of remuneration. Work may be physiological or psychic and or a combination of both; however, fundamentally it is the application of energy. An important characteristic of work is that it is a necessary aspect of socialization. Childhood can be defined in terms of age. In some societies, age may not be a sufficient basis for defining "childhood." The fulfillment of certain social rites and traditional obligations may well be important requirements in defining "adult" and "child" status. In other societies, the integration of children into socioeconomic life may begin so early that it may be virtually impossible to identify clearly the different life phases. Besides, in the absence of an effective age record system, even applying an agreed legal definition becomes highly problematic. However, in the context of child labor, a working definition of a "child" may be limited to a person below the age of fifteen years or in special circumstances fourteen years, set by the Minimum Age Convention, 1973 (No.138).[3] According to Francis Blanchard (former director of ILO) "Child Labor includes children prematurely leading adult lives, working long hours for low wages under conditions damaging to their health and to their physical and mental development, sometimes separated from their families, frequently deprived of education and training apprentices that could provide them a better future."[4] The problem of child labor is not a new phenomenon, but it was never highlighted as it is today. It is important to make a distinction between "child labor" and "child work." Child work means that the child has time for play and education besides work. Child labor, however, refers to children who are being forced to carry out the full-time work of adults. Child laborers, no matter if they work full-time in the household, in workshops, factories, or in the agricultural sector, are deprived of education, play, and a normal childhood. Child laborers can also be "bonded," which means that the children are forced to work for an employer or "master" to pay back the loan (plus often high interest) that his or her (grand)parents once received from the landlord/employer. These bonded child laborers are entirely at the mercy of the master.

The General Assembly of the United Nations adopted the Convention on the Rights of the Child (CRC) on November 20, 1989. It upholds the protection of children everywhere against exploitation, neglect, and abuse. The Convention draws attention to four sets of civil, political, social, economic, and cultural rights of every child. These include (1) The Right to Survival; (2) The Right to Protection; (3) The Right to Development; and (4) The Right to Participation. Article 23 of the Convention says, "All States should work to end such child labor practices and see how the conditions and circumstances of children in legitimate employment can be protected to provide adequate opportunity for their healthy upbringing and development."[5] The UN is conscious of the fact that the rights of the children are violated across the globe. With this backdrop a world conference was organized and sent a strong message to all those countries where children's rights are abused the most.

The World Conference on Human Rights organized by the UN in Vienna, Austria, June 14–25, 1993, reiterated the principle of "First Call for Children"[6] and underlined the importance of major national and international efforts, especially those of the United Nations Children's Emergency Fund (UNICEF), for promoting respect for the rights of the child to survival, protection, development, and participation. It called on states to integrate the Convention on the Rights of the Child into their National Action Plans.[7]

The World Conference on Human Rights urged all states, with the support of international cooperation, to address the acute problem of children under difficult circumstances. It advocated addressing the root causes of exploitation and abuse of children and for effective measures against harmful child labor. Also, the conference supported all measures by the United Nations and its specialized agencies to ensure the effective protection and promotion of human rights of the female child. It urged states to respect existing laws and regulations and remove customs and practices that discriminate against and cause harm to female children. It recognized the important role played by NGOs in the effective implementation of all human rights instruments and, in particular, the Convention on the Rights of the Child (Vienna Declaration and Programme of Action, 1993).[8]

Since as early as 1921, when the International Labor Organization (ILO) passed the first Minimum Age Convention, the world has attempted to protect children's right to an education and to prevent any child labor that would prejudice their school attendance.[9] The ILO's Minimum Age Convention 138 of 1973[10] set the standard for the minimum age for admission to employment as fifteen years, or in special cases where economic and educational facilities are insufficiently developed, fourteen years; light work not harmful to the child or prejudicial to his or her attendance at school is permissible after age twelve. Since 1990, with the entry into force of the Convention on the Rights of the Child,[11] the child's right to be protected from "any work that is likely to be hazardous or to interfere with the child's education" (Article 32) and his or her right, on an equal, non-discriminatory basis to "primary education compulsory and available free to all" (Article 28) have gained the status of internationally recognized norms, while imposing an obligation on the 192 states parties to the Convention to realize these rights for the children under their jurisdiction.[12]

The United Nations Declarations emphasized child welfare, having applications to every social condition, by conferring specific rights to children through social legislation prohibiting any discrimination.[13] The year 1979,[14] was designated as International Year of the Child (IYC) by the United Nations, which stressed the cooperation of the member nations in the common tasks of meeting the basic needs of children, including nutrition, health, education, maternal protection, family care, equal society status, and protection from racial and other forms of discrimination.[15] This was a challenge to the conscience of mankind and to the community of nations.

The ILO's Convention No. 138 of 1973 emphasized that each member for which the convention was in force must undertake to pursue a national program designed to ensure the effective abolition of child labor and to raise regressively the minimum age for admission of children to employment of work. The same convention also recommended for special attention the working conditions of young persons below eighteen years, in terms of fair remuneration, limited working hours, prohibited night work, weekly and annual holidays with leave, coverage by all social schemes, maintenance of satisfactory standards of health, safety, and so on. The ILO has adopted five conventions for making medical examination of young persons a condition precedent for employment. This is to ensure fitness of the young persons for the job where he/she is to be employed.[16]

## CHILD LABOR IN THE GLOBALIZED WORLD

The intensification of worldwide social relations through globalization has led to the changes in almost all spheres of human life. The demands on women to increase family income in the face of inflation and decreasing social-sector budgets have resulted in more children being put to work either in the domestic chores in the case of girls or to add to the family income. There is a concrete link between government policies, globalization, and liberalization and the economic, social, and infrastructural growth and development that result in child rights violations. The benefits accruing from globalization are undeniable but because of globalization's inherent structural defects, it is horizontal to conflict and social inequalities in the short term, if not synchronized and coordinated within the local context. By stringent enforcement of laws, regulations, and social tagging processes, the globalization process in diverse sectors of economy will have constructive consequences for decreases in child labor. However, without proactive measures, globalization has increased the child labor component in urban sector economies, with the greatest sufferers being girls, whose dropout rates from schools have increased.

Today, the world is in a phase of transition in terms of integration of economic activities between nations on a global scale. Developing countries compete to get a higher share in the world market for products that are in increasing demand by consumers in the developed countries. The competition seems to be based on minimizing cost of labor rather than on maximizing the skill of labor. Child labor and bad labor conditions are increasingly related to this phenomenon. Though the integration of economic activities in the world has always existed, today the process takes place on a larger scale and has to satisfy demand in a tighter time schedule. Developed countries still shift labor-intensive parts of their production to "cheap labor economies" and, to save time, they subcontract work to their local informal sector because of its fast and flexible working methods. The majority of the developing countries—Bangladesh, Pakistan, India, Sri Lanka, to mention a few—have a competitive edge on cheap bulk consumer products such as bed sheets, garments, (e.g., T-shirts, shirts, and nightwear), and shoes, but also on carpets and some food products. But, their competitive edge is based on the availability of a large, cheap, docile, and flexible workforce: mainly women and children of the poorest section of society who are employed in the informal sector where terms and conditions of employment are substandard. Exploitation of the workers is the consequence of the process where workers are often unregistered, temporary, unorganized, and earning less than minimum wage for long hours of work without the protection of social security.

The process of international integration of economic activities through liberalizing international trade has pushed more workers into the informal sector of both developing and developed countries. In the developing countries, more than in the developed countries, this has led to the economic exploitation of children in the production of goods, both for export and the domestic market. According to the ILO, more children may be involved in the urban informal sector than in the agricultural sector, because of fast-growing migration into the cities and the decentralization of production units.

The recent research shows that children are less likely to work in countries that have more international trade. The negative association between trade and child labor holds even when considering only poor countries' trade with high-income countries. It also holds up for trade in unskilled-labor-intensive products. Quite simply, child labor is less prevalent in countries that trade more because countries that trade more are richer, and children work less in richer countries. Why, then, is public perception so strongly tilted in favor of the idea that international trade causes child labor in poor countries? The fact that some children are engaged in the production of exports leads many to conclude that the availability of work in export sectors induces children to work. Yet, child involvement in the production of products for export is not evidence that the export opportunity causes children to work.

Coming to the ongoing debate on the impact of globalization and economic reform on child employment, one finds differing opinions being expressed by economists and social scientists. Some perceive that globalization increases child labor and they marshal several reasons for this. They argue that decentralized production process and intensified competition over wage costs lead to an increase in the number of employed children. Secondly, the weakening of trade unions has also resulted in the fall of real wages of workers, and in the absence of alternative gainful employment children are pushed to the workplace at a tender age. Some, on the basis of cross-country empirical investigations, argue that globalization does not necessarily increase child labor. Either it reduces or at best has no significant effect on child labor. This is because the globalization process increases the real wage rate of largely uneducated workers relative to educated workers. The increase in real wages makes poorer families less poor and enables them to finance investment in the education of their children and thereby complements the other ongoing effect, aimed at combating the problem of child labor. Thus for a developing country, with a large uneducated workforce, trade expansion seems to help in reducing child labor. The last group of opinion makers takes an eclectic position by saying that the relationship between trade and child labor depends upon the type of activities a country specializes in, and the macroeconomic policies it pursues.

## GLOBALIZATION AND DECREASE IN CHILD LABOR

A study by Alessandro Cigno, Furio C. Rosati, and Lorenzo Guarcello (1992) has indicated that there is no empirical evidence that indicates trade exposure per se increases child labor. As trade theory and household economics lead us to expect, cross-country evidence seems to indicate that trade reduces or, at worst, has no significant effect on child labor. Consistent with this theory, a comparatively well-educated labor force and active social policy appear to be conducive to a reduction in child labor. For countries with a largely uneducated workforce, the problem is not so much globalization, but mostly ineffective social mobilization and economic resources.

It indicates that exposure to international trade does not necessarily encourage child labor if policies are regulated and monitored effectively with appropriate synergy of actions from government, civil society, trade unions, entrepreneurs, and corporations. In fact, if properly harnessed, globalization should spur growth and economic development, leading to social advances including the elimination of child labor. The study has indicated that higher foreign trade between countries appears to be associated with a lower incidence of child labor.

Here again, however, we find considerable national differences. At low levels of trade, countries similar in terms of exposure to trade can be very different in terms of child labor incidence. A clearly negative association between trade openness and child labor is evident only for Africa and Latin America. In Asia, the relationship between trade and child labor is much weaker (especially if the latter is measured as a complement to the school enrolment rate). This seems to indicate that the type of activity in which a country specializes and the policies it pursues shift the relationship between child labor and globalization.

There is, therefore, no prima facie evidence that globalization will necessarily result in more child labor. Indeed, there are signs that international trade and economic integration offer governments the opportunity to reduce child labor. Since trade promotes economic growth, the opportunity could come in the form of higher income. Or it could be that the relative wage changes brought about by international trade are conducive to less child labor. Globalization can enhance employment and earnings in developing countries because of inflows of foreign investment or increases in the value of a developing country's export products. When a country opens to international markets, foreign investment, often (but not always) enters the country. This leads to increase in the demand for local labor and hence higher wages. In addition, many of today's developing countries have a comparative advantage in agriculture, and integration into international markets may increase the price of the export product to international levels. Thus, trade liberalization may increase employment and wages in these agricultural export sectors. These changes in developing country labor markets stemming from globalization could increase improvements in earning opportunities and can reduce child labor.

The argument is that the increased parental earnings may help parents reduce the work that children perform. Parents can buy substitutes for goods previously produced by children, or they can use their increased income to substitute for the money previously earned by children. A working child in the family, as a result of globalization, may provide incentives for other children in the family to stop working and attend school. Hence, globalization can help parents in poor countries to stop child labor without foreign intervention. A study conducted in Vietnam has suggested that when globalization improves the income of impoverished households, the additional income helps parents reduce the workload of their children and allows them to send more of them to school.

## GLOBALIZATION AND INCREASE IN CHILD LABOR

There are other viewpoints, however, that indicate that globalization is out of control and instead of reducing child labor it actually increases child labor, mainly in export industries. Although large-scale manufacturing export industries in South and East Asian countries do not directly rely on child labor, there are linkages to child labor, through subcontracting, in the labor-intensive, low-wage segments of the production of some export items such as carpets, garments, rattan furniture, sports products (especially soccer balls), and precious stones notably in India, Bangladesh, Pakistan, Thailand, Philippines, and Indonesia. In 1999 and 2000, UNICEF interviewed children across thirty developing countries and discovered that factory work is rare. Only 3 percent of children worked outside their household for payment. Unpaid domestic work is more common than work for payment. UNICEF found that 2 percent of boys and 5 percent of girls perform unpaid domestic work for households other than their own. However, most working children are employed in their own households, helping with household duties or the family's farm or business.

It is also true that globalization increases a country's exposure to foreign competition. This may force inefficient firms in import-competing industries out of business. Regardless of the long-term benefits of this reallocation of resources, in the short term, these adjustments may create difficulty for some households with children.

In such export industries child labor has little to do with poverty, culture, or tradition. It is much more the result of greed, connivance, and complacency. It is not at all surprising, especially when a child is paid only one-sixth to one-third of the adult wage. Governments turn a blind eye to the practice and in many cases enter into agreements with investors, so that the labor laws of the country are not applied. Thus the complacency of the market (i.e., all of us) in its desire for cheaper consumer products, regardless of the exploitation involved in production, is the main cause of increase in child labor.

Social scientists differ in how they view a parent's decision to send a child to work. The debate revolves around how parents in developing countries view child labor. In one view, parents want their children to work. Perhaps parents even have children in order to send the children to work. This view implies that parents take advantage of any earning opportunity. So long as parents benefit more when the child works than when the child goes to school, child labor will persist. Improvements in living standards may do nothing to ameliorate child labor. In such situations globalization is bound to increase child labor if it is not regulated and monitored by the government and civil society organizations.

The effects of globalization on child labor growth due to the exports of cotton from Uzbekistan; carpets from India; garment manufacturing from Bangladesh, India, and Honduras; and sports products, especially soccer balls from Pakistan are classic examples of linkages between globalization and child labor. This has been repeated all around the world with children being abused in the most horrendous way.

Households that cannot meet their basic needs depend upon the income of their children for survival. Without the income from working children, parents in poor households may have to choose which child to feed. Moreover, schooling is expensive in most of the developing countries. Thus proactive measures of enforcing child labor laws by national governments, supported by provisions for social and economic security to the families with support from international organizations, business houses, and national governments is likely to go a long way in eliminating child labor even if the globalization process exists.

Unlike in the case of nationally consumed products, the silver lining in the case of export-oriented industries—due to the rapid process of globalization—is that consumers across the world no longer find the abuse of children acceptable. They no longer want to be part of this child exploitation. There has been an awareness generated and people are refusing to buy goods identified as having been made by children. The momentum generated by social labeling has pressurized international business houses to introduce ethical codes of practice and also to support international organizations and national governments financially in order to strengthen effective measures for protection of children.

This has nothing to do with morals and everything to do with economics. A recent study suggests that up to one-third of a company's share value is dependent on its image. A reputation for merchandising or retailing items made by children is not good for that image. Such social labeling has pressurized governments to make appropriate legislation against child labor, accept international conventions on child labor exploitation and child protection, and implement these laws and international conventions earnestly. Preservation of their export market has served as an effective incentive to the carpet, garment, and other export producers in India, Pakistan, and Bangladesh respectively to actively support IPEC action programs to withdraw and rehabilitate child workers engaged in these industries.

Child labor is a universal problem. It exists all over the world, but it is concentrated (98 percent) in the developing countries. In the countries where child labor exists, it is often stated that poverty is the main reason why children are sent to work and that poverty first needs to be solved to eradicate child labor. But, as experts state, "child labor also perpetuates poverty as children become part of the destructive inter-generational cycle of repetitive impoverishment," "child labor adds to unemployment of adults as they take the jobs of as many unemployed adults, reduces the need for technological innovations which is so essential to the expansion of exports," "holds wages down and, hence, increases the number of families living below the poverty line." Thus, two important causes of child labor as stated by many labor experts are the vested interests of employers in employing cheap child laborers and the inefficiency and inadequacy of existing primary education facilities. The first cause implicates the interest of employers, who can pay low wages, extract more work, make maximum profit, and remain free of any labor unrest. Child laborers are also victims of physical, mental and sexual abuse by employers. The second cause implicates government schools that are poorly run and maintained and lack teaching materials and motivated teachers, causing children to lose interest in education.

## CHILD LABOR AND DEVELOPING WORLD

A study conducted by the ILO Bureau of Statistics found that "Children's work was considered essential to maintaining the economic level of households, either in the form of work for wages, of help in household enterprises or of household chores in order to free adult household members for economic activity elsewhere." The question of how trade liberalization affects the employment of children in developing economies in the process of globalization all over the world is a subject of debate. The opponents of globalization argue that market integration, by increasing labor demand, expands the earning opportunity of children. And thereby it inevitably leads to more child labor. Some even propose the use of trade sanctions against countries where a high level of child labor has been used.

For the developing countries, where the vast majority of the world's working children are to be found, adaptation to the globalizing economy basically means two things. First, they must be able to expand their exports, and second, they must be able to attract a greater volume of direct foreign investment. But a survey conducted by the ILO on the handwoven carpet industry in India suggests that the savings by consumers in developed countries through the employment of children was a mere 5–10 percent depending on the type of carpet. And there was nothing to indicate that countries with poor labor standards achieve better export results globally than countries with high labor standards.

Child labor impedes children's intellectual development and prevents them from becoming productive adults. The prevalence of child labor is already creating obstacles in developing countries' exports. It is claimed that globalization will be beneficial in the long run. Debates on child labor and its relationship with the process of globalization and structural reforms are a fairly recent development. As India accounts for a substantial contribution to the global child labor pool, there is an emergent need to understand the complex relationship between child labor, exploitation of working children, violation of economic and social rights of children, and the ongoing process of globalization.

Austerity policies in the form of reductions in social-sector expenditure and the introduction of user fees intended to stabilize the macro economy have a strong link with child labor and rising dropout rates of children. These policies directly affect poor households, and in particular women and children, by raising the cost of living for families. To finance the increasing costs and in order to generate the additional income women are pushed to join the labor market in low-paid jobs. Children are compelled to leave schools and contribute to the family income at an earlier age than before. A common pattern is that the girl child is the main loser when school fees are introduced or when prolonged unemployment forces families to withdraw children from school. Further structural adjustment polices change the relative prices between paid and unpaid work in favor of the former. The invisible work done by the children and women is not counted.

## CHILD LABOR AT THE GLOBAL LEVEL

Child labor is a pervasive problem throughout the world. Industrialized economies—Europe, North America, Australia, and others—have by now reduced employment of children within their borders to a considerable extent, but they are yet not fully out of the phenomenon of child labor. Some of them have some child labor originating from social problems, poor educational systems, and poverty. But the problem of child labor as faced by developing economies today has indeed serious dimensions. Africa and Asia together account for over 90 percent of total child employment. Though there are more child workers in Asia than anywhere else, a higher percent of African children participate in the labor force. Children in developing countries also start work at a younger age, sometimes but not always within the family. Moreover, the variety of jobs these children do is greater than in the developed countries, and less recognized by the authorities. They enjoy fewer legal or other protections, if any, and receive little or no training to help them deal with the health and other hazards of their work. They are often helpless to counter poor treatment and exploitation by their employers (or by their families). All in all, children in developing countries work much harder than those in industrialized countries, for less reward, and most often forgo the benefits of schooling. Some even have to create their own jobs, which they do by working on the streets or scavenging for garbage.

Young workers in the developed countries are in a better position. Perhaps their biggest advantage is that many of them manage to combine work with schooling, and one reason for this is the heavy emphasis on school attendance by the local authorities. Africa has the highest proportion in the world of working children (nearly one-third), while Latin America, with its high levels of urbanization, has the largest population of "street children." And in many Asian countries children comprise over 10 percent of the work force. Yet child labor is remarkably persistent and remains widespread in much of the developing world, including in the booming parts of the world economy. A 2003 survey by the ILO suggested that there were 246 million child laborers (aged fourteen years or less) in the world and that as many as 180 million of them were engaged in hazardous activities that put them at direct physical risk.

India has the largest number of child laborers in the world. The United Nations Children's Fund (UNICEF) estimates that there are more than thirty-five million such children, accounting for 14 percent of the children in the five to fourteen years age group. Other unofficial estimates are much higher, ranging between 60 and 125 million child laborers. Meanwhile, the Census of India data for 2001 suggests a much lower incidence, with 12.5 million child laborers identified, which would be less than 5 percent of the relevant age group. Out of the total population of child laborers in India females constitute 46.26 percent, whereas male child laborers made it 53.74 percent.

## TRENDS IN CHILDREN'S WORK

Millions of children work to help their families in ways that are either harmful or exploitative.[17] (See table 1.1, which shows the incidence of work among children.) But one in six children aged five to fourteen, about 16 percent of all children in this age group, is involved in child labor in developing countries. In the least-developed countries, 30 percent of all children are engaged in child labor. These children are put to work in ways that drain childhood of joy, crush their right to normal physical and mental development, and often interfere with their education.

Around one in three children aged five to fourteen in Africa labors, compared to only one in twenty in the Central and Eastern European/Commonwealth of Independent States (CEE/CIS) region. Children living in the poorest households and in rural areas are most likely to be involved in child labor. Boys are more likely to be engaged in child labor than girls because the former are more likely to be engaged in economic activity. Those burdened with household chores are overwhelmingly girls.

**Table 1.1**

**Table 1.1. Estimates of Various Forms of Children's Work, 2000 and 2004**

| Age Group | | Child Population | | Working Children | | Child Laborers | | Children in Hazardous Work | |
|---|---|---|---|---|---|---|---|---|---|
| **2000** | **2004** | **2000** | **2004** | **2000** | **2004** | **2000** | **2004** | | |
| **5–17** | **No. (millions)** | 1531.4 | 1566.3 | 351.9 | 317.4 | 245.5 | 217.7 | 170.5 | 126.3 |
| | Incidence (Percent of age group) | 100.0 | 100.0 | 23.0 | 20.3 | 16.0 | 13.9 | 11.1 | 8.1 |
| | Percent change from 2000 to 2004 | – | 2.3 | – | -9.8 | – | -11.3 | – | -25.9 |
| **5–14** | No. (million) | 1199.4 | 1206.5 | 211.0 | 190.7 | 186.3 | 165.8 | 111.3 | 74.4 |
| | Incidence (Percent of age group) | 100.0 | 100.0 | 17.6 | 15.8 | 15.5 | 13.7 | 9.3 | 6.2 |
| | Percent change from 2000 to 2004 | – | 0.6 | – | -9.6 | – | -11.0 | – | -33.2 |
| **15–17** | No. (million) | 332.0 | 359.8 | 140.9 | 126.7 | 59.2 | 51.9 | 59.2 | 51.9 |
| | Incidence (Percent of age group) | 100.0 | 100.0 | 42.4 | 35.2 | 17.8 | 14.4 | 17.8 | 14.4 |
| | Percent change from 2000 to 2004 | – | 8.4 | – | -10.1 | – | -12.3 | – | -12.3 |

## SECTORAL DISTRIBUTION OF CHILDREN'S WORK

The indicator for employment by sector—based on the International Standard Industrial Classification of All Economic Activities, Revisions 2 (1968) and 3 (1990)—breaks employment down into three broad groupings of economic activity: agriculture, industry, and services.

- The agricultural sector comprises activities in agriculture, hunting, forestry, and fishing. Children's work is primarily concentrated in agriculture—agriculture accounts for more than two-thirds (69 percent) of all working children in the age group five to fourteen.

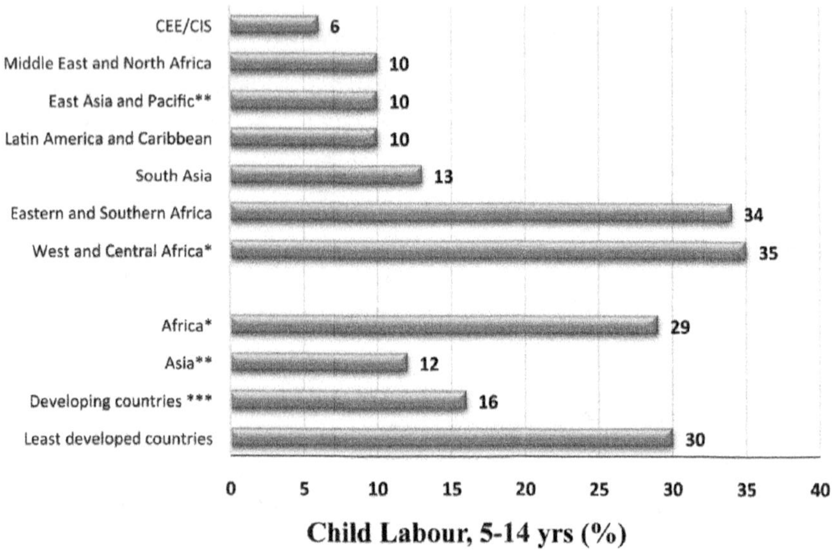


**Figure 1.1.**

- The industry sector comprises mining and quarrying, manufacturing, construction, and public utilities (electricity, gas, and water). Industry accounts for 9 percent of working children.
- The services sector consists of wholesale and retail trade; restaurants and hotels; transport, storage, and communications; finance, insurance, real estate, and business services; and community as well as social personal services. Services account for 22 percent of working children.

## UZBEKISTAN

Uzbekistan is located in the heart of Central Asia, and lies between two "sister rivers," the Amu Darya (Oxus) and the Syr Darya (Jaxartes), the largest and deepest rivers in Central Asia. The republic lies along a northwest to southeast axis, and its eastern extremity, the Fergana Valley region, borders Kyrgyzstan to the east, Tajikistan to the southeast, Afghanistan to the south, and Turkmenistan to the southwest and the west. The northwest of the country consists of the Aral Sea, and beyond that lies Kazakhstan, which forms the entire northern border. Uzbekistan covers an area of 447, 400 square kilometers, of which the Autonomous Republic of Karakalpakistan covers 164,900 square kilometers. The republic is divided into twelve provinces and subdivided into 162 districts, in which there are 118 cities, 116 towns, and 1,332 villages. The entire territory spans 930 kilometers from north to south and 1,425 kilometers from east to west. Most of the territory of Uzbekistan is flat country and desert, including the southwestern part of the Kyzl Kum (Red Sands). The climate of Uzbekistan is continental, with low precipitation, relatively high temperatures and a large number of cloudless days during the year. The population of Uzbekistan is young, with more than 35 percent under sixteen years of age. Over half of the population is female, and the life expectancy of women is seventy-two years, exceeding that of men by six years. After the creation of the former Uzbek Soviet Socialist Republic, the then leadership was forced to abandon the cultivation of grain and other food products and concentrate on providing cotton for Russia's textile industry, and thus accept a cotton monoculture.

### *The Karakalpak Autonomous Republic*

The Karakalpak Autonomous Republic was first created as an autonomous region within Kazakhstan on February 16, 1925. On March 20, 1932, it was upgraded to the Karakalpak Soviet Socialist Autonomous Republic, and from 1930 to 1936 it remained, with the same status, within the Russian Soviet Federal Socialist Republic. On December 5, 1936, the Karakalpak Autonomous Republic was transferred to the Uzbek Soviet Socialist Republic. Karakalpakistan is situated in the northwestern end of the republic of Uzbekistan. It has fifteen districts, twelve cities, sixteen towns, and 107 villages. Karakalpakistan declared its sovereignty on December 14, 1990. According to the constitution the sovereign republic of Karakalpakistan is part of the republic of Uzbekistan and the security and integrity of Karakalpakistan is protected by the republic of Uzbekistan. The laws of the republic of Uzbekistan are binding in the territory of Karakalpakistan. The economy of the autonomous republic is dominated by agriculture, mainly the cultivation of cotton and wheat. The life of the people, however, is deteriorating as a result of the environmental problems caused by the decline of the Aral Sea. The cultivation of cotton requires huge amounts of water and labor, especially during the harvest season. The extensive cotton cultivation has heavily drained water from the Aral Sea, which is now squeezed to 15 percent of its original size, and this has resulted in the death and destruction of its flora and fauna species. In addition, to meet the labor required for its cotton cultivation the authorities have been using schoolchildren time and again as a cheap labor force.

## CHILD LABOR AND COTTON INDUSTRY OF UZBEKISTAN

Uzbekistan in 1991 emerged as a sovereign country after more than a century of Russian rule—first as part of the Russian empire and then as a component of the Soviet Union. Positioned on the ancient Great Silk Road between Europe and Asia, majestic cities such as Bukhara and Samarkand, famed for their architectural opulence, once flourished as trade and cultural centers. Uzbekistan is the most populous Central Asian country and has the largest armed forces. It is one of the world's biggest producers of cotton and is rich in natural resources, including oil, natural gas, and gold. However, rigid political control is mirrored in the tightly centralized planning of the economy. Economic reform has been painfully slow and poverty and unemployment are widespread. According to an ILO report, children of school age in Uzbekistan are working in different sectors of the economy. Large numbers of school-age boys are working as potters. Young girls from the countryside are also sent to cities to work as domestic helpers, as the money they earn is a lifeline for their families.

In all regions of Uzbekistan, government officials mobilize children in an attempt to ensure that state cotton quotas are met. Schools are closed down, and children as young as seven are sent to the fields to pick cotton by hand. Headmasters are given quotas that dictate how much each student is to harvest. And those who fail to meet their targets, or who pick a low-quality crop, are reportedly punished with detentions and told that their grades will suffer. Children who run away from the cotton fields, or who refuse to take part, can face expulsion.

Statistics on children employed in the cotton sector in Uzbekistan are difficult to obtain, but the London-based rights group Environmental Justice Foundation (EJF) says around two hundred thousand children work in the major cotton-producing region, Fergana.[18] It is difficult to quantify the number of children involved. One estimate has been provided by UNICEF, which suggests that 22.6 percent of children ages five to fourteen years in Uzbekistan were working in 2000. Habib Mamatov, an official responsible for the cotton harvest of Kashkadarya region, in a public interview with Tribune-Uz, stated that 39,656 university and college students as well as 44,385 high school and middle school students were involved in the 2004 cotton harvest campaign. Extrapolating from this figure, a rough estimate for the whole country suggests that over one million children, a third of them younger than fifteen, are recruited to pick cotton each year. For their arduous work children are paid very little or nothing. According to EJF, some children working in the Fergana region stated that they worked from 7 AM until 5 PM in return for eight to sixteen cents US. Moreover, although these figures represent the official wages, in reality many youths receive no actual cash at all. Students are assessed the cost of their meals, which in practice may leave them in debt by the end of the harvest season.

In Soviet times up to two-thirds of Uzbekistan's cotton was harvested by machine. Nowadays this figure has dropped to just 10 percent. Instead, the majority of Uzbek cotton is gathered by hand, often by children. According to reports from nine of Uzbekistan's twelve territorial units (Jizzakh, Fergana, Namangan, Syr Darya, Surkhandaria, Bukhara, Khorezm, Tashkent, and Samarkand provinces) by the third week of September local governments and school administrators sent children as young as the seventh grade (ages thirteen to fourteen), and in some cases as young as fifth grade (eleven to twelve) out to the fields to pick cotton. By the end of September, pressure to bring in the harvest before rains led local officials to order the smallest schoolchildren, from first grade on, to labor on the harvest. In Fergana, schools are closed and children are sent out in September, though a week earlier children are forced to sign statements that they will remain in school after the end of the semester. Experts suggested that the statements are intended to give local government officials plausible deniability if the children's presence in the fields is challenged.[19]

In Namangan, human rights defenders observed children from several schools, some as young as eleven, picking cotton. Every day local government officials and bureaucrats from the local education department visit the fields to check up on the number of pupils out picking, and to make sure that harvest targets are being met. Similarly the Samarkand provincial government also sent its schoolchildren out to pick cotton in September. Children as young as thirteen are forced from their classrooms, though high school, junior college, as well as university students (ages sixteen and above) are also sent out to the fields for several weeks.[20]

Children recruited to pick cotton in their vicinity are able to return home to their families in the evenings. But older children and those conscripted to work in the more remote cotton farms are forced to sleep in makeshift dormitories on farms or, ironically, in classrooms, often with poor living conditions, at times drinking irrigation water and with insufficient or poor-quality food. Some children recount living in barracks with no electricity, windows, or doors. After weeks of arduous work and poor accommodation children can be left exhausted and in poor health. Human Rights Organizations confirmed eight deaths of children working in the Samarkand region over a two-year period.[21] Many more suffer with chronic diseases including intestinal infections, respiratory infections, meningitis and hepatitis.

While it is certain that the Uzbek regime exploits children in forcing them to take part in the annual cotton harvest, it is less clear how much the children are paid. Some claim to receive US $5 for five days' work. Others report receiving just 15 US cents for the same period of labor. In 2001, the Uzbek NGO Tahlil estimated that payment for 1 kg of cotton ranged from 1.5 US cents at the beginning of the season to 1.0 US cent at the end. In 2004, children in the Fergana region reported that an average day's harvest of 10 kg of cotton would earn them 38 US cents. For the Uzbek regime, forced child labor is undeniably cheap and immensely profitable. A child may be paid, at best, 3 or 4 US cents per kg for a commodity that is valued at US $1.15 on the global marketplace. Once the cotton harvesting begins many schools are closed down as children, some as young as ten, are sent to the fields to pick cotton by hand for up to three months. They receive little, if any, pay.

Under the Soviet Union, forced labor was accompanied by some care for the health of the children, the quality of their nutrition, and development of the rural social infrastructure. The Paris-based group Human Rights in Central Asia reported, "Now forced labor is compensated neither by decent payment, nor through public funds. Every year, starting September, schools across the country are closed for more than two months. Students are forced to pick cotton by order of central and local authorities. Children work at least eight hours daily on the cotton fields, sometimes without rest for days. They inhale dust, laden with residues of chemicals, pesticides and defoliants used in the fields before the cotton harvest."

The use of children ensures maximum profits to the ruling elite, which benefits from the export of cotton to Western consumers. The use of child labor violates international laws and conventions to which many governments of cotton-producing countries are signatories. Children's normal education is interrupted to serve the interest of the small elite who benefit hugely from the high profits from trading cotton on the world market, reported Initiative Group of Independent Human Rights Defenders of Uzbekistan based in Tashkent. As a result of forced child labor, children cannot attend schools and colleges during the academic period, and they lag behind in the school curriculum, while some children fall sick from hard work and exhaustion.

Refusing to collect cotton can be punished by expulsion from the educational institution. The students are also beaten up by school staff for refusing to work for the cotton harvest. The tens of thousands of children forced to work in the fields each year provide more than half of the cotton produced in Uzbekistan. The government's Soviet-style cotton quotas and unwillingness to pay a decent living wage ensure that children are used instead of adults.

The UN convention on the rights of a child provides that children have a right "to be protected from performing any work that is likely to be hazardous, or to interfere with the child's education, or to be harmful to the child's health or physical, mental, spiritual, moral or social development."As forced labor on the cotton harvest prevents children from attending school for over a third of the academic year, it clearly violates the children's right to an education. Work on the harvest and exposure to pesticides and defoliants is also demonstrably detrimental to children's health, Human Rights Watch (2001) in Central Asia reported.[22] Cheap clothing and other cotton items in the developed world are being subsidized by child laborers in poor cotton-producing countries.

## COTTON MONOCULTURE: HISTORICAL BACKGROUND

Uzbekistan's cotton monoculture is a legacy of Soviet rule. The centrally planned Soviet economy aimed to make the USSR self-sufficient. To this end certain republics became highly specialized producers of certain commodities for consumption within the Soviet market and in certain cases to export for hard currency. The Soviet regime concentrated on growing cotton in Uzbekistan at the expense of all other crops. As a result of this cotton monoculture Uzbekistan continues to suffer from a variety of economic, political and ecological deformities today. Not the least of which has been the drying up of the Aral Sea and the poisoning of the surrounding land with salt.

Already in the 1860s in response to the loss of the Southern United States as a source of cotton, the tsarist regime sought to promote the cultivation of cotton in what is now Uzbekistan to provide raw material for its textile mills. In 1925 and 1926, Soviet land reform eliminated the problem of landless peasants cultivating other people's lands in the region. However, most of the small plots owned by Uzbeks remained too small to be economically viable. The inability of Uzbek farmers to make a living from these parcels of land facilitated the collectivization of agriculture in the early 1930s. Unlike European areas of the USSR, strong class antagonisms between the formerly landless peasants and the former landowning class, or *bais*, remained muted. Many *bais* retained positions of notable influence in rural communities despite the confiscation of much of their wealth during the land reform campaign. To remove this influence, which the Soviet government correctly connected with opposition to collectivization, the OGPU forcibly deported over twelve thousand families from their homes to places as far away as Ukraine and the North Caucasus. The collectivization of agriculture ensured state control over the rural Uzbek economy, and in the following decades, the Soviet state used this control to turn the Uzbek countryside into one large cotton plantation.

The Soviet government sought to expand the production of cotton in Uzbekistan by expanding its planted acreage at the expense of food crops and by the intensive use of fertilizers. Expansion of cotton cultivation required massive irrigation. Unpaid corvée labor built the canals that provided this irrigation, including the 270-kilometer-long Great Fergana Canal constructed in 1930. Other forms of forced labor also contributed to cotton cultivation at this time. By 1934, over twenty thousand inmates in the Sazlag complex of corrective labor camps in Chirchik worked on cotton farms. By the end of the 1930s, the USSR had become self-sufficient in cotton.

Cotton cultivation in Uzbekistan expanded massively between 1960 and 1980, when the area under irrigation increased from 2.3 million to more than 4 million hectares. More than 2 million hectares of irrigated land were planted with cotton during the 1980s, which shipped to other Soviet republics for processing. Since independence, cotton exports have increasingly shifted to Western markets, including the United States and Europe. In recent years Uzbekistan has been the world's fifth-largest cotton producer (with a world market share of 6 percent) and the second largest exporter (17 percent of world exports) (table 1.2). Cotton export earnings peaked in 1995 at US$1.6 billion, significantly boosted by high world market prices. Although cotton's share in total exports declined to less than 40 percent in 1997, the "white gold" remained by far the single largest source of foreign exchange for the economy.

In the past few years the area under cotton has been lowered to 1.5 million hectares, which in part explains the decline in cotton output observed in recent years. A second important factor in 1996 and 1997 was

**Table 1.2**

**Table 1.2. Cotton World Production and Exports, 1992/93–1997/98**

| | 1992/93 | 1993/94 | 1994/95 | 1995/96 | 1996/97 | 1997/98 |
|---|---|---|---|---|---|---|
| World Production | 17,9611 | 16,707 | 18,639 | 20,254 | 19,417 | 19,296 |
| *Of Which* | | | | | | |
| United States | 3,531 | 3,513 | 4,281 | 3,897 | 4,124 | 4,092 |
| China | 4,507 | 3,745 | 4,333 | 4,768 | 4,202 | 4,289 |
| India | 2,346 | 2,066 | 2,354 | 2,885 | 3,000 | 2,439 |
| Pakistan | 1,540 | 1,368 | 1,361 | 1,785 | 1,589 | 1,524 |
| **Uzbekistan** | **1,274** | **1,321** | **1,254** | **1,250** | **1,034** | **1,176** |
| Turkey | 574 | 602 | 628 | 852 | 784 | 718 |
| World Exports | 5,575 | 5,815 | 6,176 | 6,065 | 5,768 | 5,687 |
| *Of Which* | | | | | | |
| United States | 1,132 | 1,494 | 2,047 | 1,671 | 1,495 | 1,633 |
| **Uzbekistan** | **1,197** | **1,263** | **1,090** | **985** | **991** | **936** |
| Franc-Zone Africa | 446 | 441 | 584 | 609 | 735 | 784 |
| Australia | 369 | 366 | 293 | 319 | 533 | 555 |
| India | 234 | 66 | 18 | 145 | 277 | 27 |
| Pakistan | 256 | 69 | 32 | 312 | 26 | 76 |
| Argentina | 65 | 69 | 197 | 266 | 290 | 218 |
| Memorandum Items | | | | | | |
| **Uzbekistan/World Production (%)** | **7** | **8** | **7** | **6** | **5** | **6** |
| **Uzbekistan/World exports (%)** | **21** | **22** | **18** | **16** | **17** | **16** |

the reduction in yields, reflecting the distorted incentive structure for cotton growers, including low remuneration for cotton pickers, poor agricultural practices, forced labor of schoolchildren in cotton cultivation, and environmental degradation. Although Uzbek cotton yield is still higher than in other Central Asian countries, it is below China's and that of other countries where cotton is grown under similar conditions.

Cotton cultivation is very labor intensive. Under the Soviet regime, picking cotton was generally undertaken during a two- to three-month campaign with the forced participation of students and teachers in exchange for very low remuneration. This practice—which involves child labor—still exists on a large scale in Uzbekistan, partly because more and more agricultural machinery has fallen into disrepair. Since independence, agricultural wages have remained very low in US dollar terms, and they have fallen sharply relative to those in industry and other sectors of the economy. Moreover, agricultural workers are often compensated partly in kind, while cash payments are delayed. Although a labor shortage exists during the peak cotton harvesting season, there is otherwise considerable hidden underemployment in agriculture (equivalent to almost 10 percent of the total number of employed persons in the country). The creation of more employment opportunities especially in the rural areas is of crucial importance, as about 40 percent of the rural population is under eighteen years of age. At the aggregate sectoral level, labor productivity has declined since 1991, as overall output has fallen and the overall level of employment in the sector, as officially measured, has remained broadly the same.

Heavy use of fertilizers and pesticides in large-scale cotton irrigation has resulted in salinization, decreasing soil fertility, and causing falling area yields. Water use in cotton cultivation is excessive due to poor water management and low or unenforced water fees. Cotton growers in various parts of Uzbekistan use three to four times, as much water as in other countries. The massive diversion of water from the two major rivers (the Syr Darya and the Amu Darya) for cotton irrigation has contributed to the sharp decline in the supply of water to the Aral Sea, which is drying up. Once the world's fourth-largest inland sea, it has lost about one-third of its surface area since the 1960s. In addition, the excessive irrigation water that is drained off irrigated land is heavily contaminated with mineral salts and fertilizers, polluting what remains of the lake. The fishing industry has collapsed, and human health has also been adversely affected as both the supply and quality of drinking water has worsened. Owing to air pollution through wind-borne salts, there has been a large increase in respiratory and other diseases among the population around the Aral Sea. The natural climate around the Aral Sea has also changed.

Cotton cultivation in Uzbekistan continued to receive a priority throughout the Soviet period. It also created significant distortions in the region's economy. By the early 1980s the USSR had become a major exporter of cotton. It accounted for over a fifth of the world's production and lagged only behind China in total output. Although Tajikistan and Turkmenistan contributed to Soviet cotton production, Uzbekistan remained the center of cultivation in the USSR with 70 percent of production. Uzbekistan became heavily dependent upon harvesting raw cotton for the USSR and had to acquire almost everything else including textiles from other regions of the USSR. Cotton in Uzbekistan at this time employed 40 percent of its total labor force and generated 65 percent of the republic's total economic output. The rapid expansion of this crop, however, proved unsustainable. It could not grow fast enough to fully employ and provide a rising standard of living to the increasing Uzbek population. The economic conditions of the predominantly rural Uzbek population thus stagnated and started to decline, a trend that accelerated rapidly after the collapse of the USSR. The Soviet emphasis on the production of raw cotton to the exclusion of other economic activities impeded economic diversification, industrialization, and urbanization. Thus Uzbekistan was left relatively backward compared to Russia and other European areas of the USSR.

Cotton monoculture also wreaked havoc on the Uzbek ecology. The expansion of cotton cultivation in Uzbekistan required the massive diversion of scarce water resources for irrigation. Cotton farms drained so much water from the Amu Darya and Sir Darya rivers that they could no longer replace the water that evaporated from the Aral Sea. As a result the Aral Sea shrank significantly and rendered much of the nearby land infertile due to excess salinity. The overuse of pesticides also posed environmental dangers. Finally, the failure to practice crop rotation led to massive soil erosion in Uzbekistan. The environmental degradation caused by cotton in Uzbekistan has caused serious health problems among many of its rural inhabitants. Despite being a noticeable problem since the 1980s, neither the Soviet government nor the Islam Karimov regime made any serious attempt to address its root causes in the cotton industry.

Today the cotton farms remain state property. The workers on these farms receive only a small fraction of the money earned by the state by the sale of cotton to Western companies. It returns very little of this to the people of Uzbekistan in the form of social services such as education and health care. Both of which have deteriorated significantly since the collapse of the USSR. Child labor mobilized by local officials during the harvest season remains widespread. This practice has further retarded education in Uzbekistan. The cotton monoculture of Uzbekistan remains in need of serious reform. Toothless measures have not improved the situation. Only the threat of sanctions against the crop unless there are serious reforms is likely to improve the situation

## COMPARISON WITH OTHER POST-SOVIET CENTRAL ASIAN COUNTRIES

Child labor is prevalent in almost all Central Asian States, but in different ways and to different degrees. Tajikistan uses child labor as a "lifeline for their families," who are forced by extreme poverty to take their children out of school and put them to work on the family farm or marketplace. As is common in all Central Asian States, the Tajik government makes pronouncements against the practice, but does nearly nothing to curb its actual use. Tajikistan has a relatively young and rapidly growing population, with 48 percent under eighteen years of age. Most families in Tajikistan have many children. The growing economic hardship has led families to increasingly count on the income derived from children's informal work, mainly performed in the street after school hours. Children usually work in the market places, streets, and other public places, which increases their vulnerability to prostitution and trafficking. According to unofficial estimates, 45–55 percent of children from the ages of ten to fourteen, especially from low-income families, are engaged in physical labor, and this percent is increasing in Dushanbe and other large cities in Tajikistan. The research done by different agencies including Asian Development Bank, UNICEF, UNESCO, and others shows that children are being heavily exploited in terms of wages, work, working conditions, and health conditions. According to the report, the Tajik legislation contains provisions on the minimum employment age, as well as the protection of children under eighteen from labor conditions that might endanger their health, safety, or morals, in line with the international legal standards. Nonetheless, in reality, there is a clear disparity between the law and the practice.

The Kazakh legislation contains provisions on the employment of minors, such as the age limit, safety measures, and various safeguards, although many children work illegally, particularly in the rural areas. The types of labor performed by children include subsistence farming, work in private enterprises, family businesses, and others. One of the main reasons for the employment of children is the low living standards of the population. The socio-demographical and economic situation in Kazakhstan is conducive to the use of cheap child labor, both during the agricultural season, irrespective of the educational process, and by family and criminal business operations. Analysis of the legal instruments of the Republic of Kazakhstan relevant to child labor shows that, irrespective of the magnitude of the problem and the significant number of both domestic and international legal provisions in force, children's rights (including labor ones) are not properly protected.[23]

Since the collapse of the Soviet Union, the use of child labor in Kyrgyzstan has become widespread, especially in the southern regions, where tobacco, rice, and cotton are cultivated.[24] Furthermore, the labor of children is traditionally used in irrigating, weeding, and harvesting work. Almost all the children living in rural areas work on plantations and help their parents in performing all types of labor, including those that may pose health hazards. The use of children in markets has become frequent due to the emergence and growth of unregulated trading activity and markets. Children are involved in transporting (unloading) and the sale of goods and luggage on the streets and in the markets, in addition to collecting bottles and aluminum. There are also children involved in tobacco-growing work, as house servants—cleaning, washing, and so on—gathering firewood, and working in gardens. Kyrgyzstan's general unemployment situation fosters child labor, since it is easier to handle children than adults. They are unaware about their rights, cause fewer problems, complain less, and are more adaptable. They are also the easiest to sacrifice when difficulties arise. Child labor in Kyrgyzstan is largely the result of cultural, historical, social, and economic conditions, where children have to work along with their parents and siblings in order to contribute to the family income, starting with the age of five.

An overview of the Turkmen legislation of labor rights of minors shows that the state is party to the main international legal documents regulating this issue. Provisions on labor performed by minors are contained in Article 179, which stipulates that persons sixteen years of age and under are not allowed to enter a labor contract. Yet pupils including those from junior grades, have to work, throughout the cotton picking season in Turkmenistan. Child labor is used for weeding during summer holidays. Many families cannot provide for their children, due to lack of employment or wage arrears. Thus begging, prostitution, and various crimes have become widespread among children, as a result of scant payment for their work, their parents' poverty and lack of education.[25] Due to stringent employment rules and regulations in Turkmenistan the problem of unemployment is expected to continue due to limited development of the private sector.

The use of child labor in Uzbekistan has largely stemmed from the difficulties of the post-Soviet transition period, which have been conducive to the use of children's labor. As a consequence of the deteriorating economic conditions, the traditionally large families have not been able to cover the education expenses of all children. The rising unemployment has reduced the levels of average income, especially in rural areas, where employment has a largely seasonal character. Poor and socially vulnerable families have not been able to maintain an appropriate quality of life.

The economics of Central Asian cotton is simple and exploitative. Millions of the rural people work for little or no reward for growing and harvesting the crop. The considerable profits go either to the state or small elites with powerful political ties. Forced child labor and other abuses are common. The industry relies on cheap labor. Schoolchildren are still regularly required to spend up to two months in the cotton fields in Uzbekistan. Despite official denials, child labor is still in use in Tajikistan and Turkmenistan. Students in all three countries are forced to miss their classes to pick cotton.[26] Little attention is paid to the conditions in which children and students work. Every year some fall ill or die.

Women also do much of the hard manual labor in cotton fields, and reap almost none of the benefits. Cash wages are minimal and often paid late or not at all. In most cotton-producing areas, growers are among the poorest elements in society. Not surprisingly, young men do everything to escape the cotton farms, forming a wave of migrants both to the cities and out of the region.

On the murky underside of Central Asia, there are underage prostitutes for sale on the streets with few rights and fewer opportunities. Child prostitutes may be virtually invisible in the Central Asian Republics, but they are there in discreet clubs, in private homes converted into brothels, and hanging around on street corners. The teenage girls of Kyrgyzstan, Kazakhstan, Uzbekistan, and Tajikistan are bought and sold as commodities, and in some cases shipped off to become sex slaves in the Gulf. A high premium is placed on virginity. Some of the worst cases involve parents selling their own daughters for gain or out of sheer desperation. Mostly the girls are aged between eleven and sixteen—although many start earlier. Some boys are involved, too. These adolescent children are very much the victims of the tumultuous changes these countries have undergone since the collapse of the Soviet Union. Many come from impoverished rural families left unable to cope by years of economic decline. Others, from broken homes or abusive family environments, have fallen through a social safety-net worn thin by lack of government spending.[27]

Sex before the age of consent (sixteen) in Central Asia is illegal, and any adult involved with a minor would face a lengthy spell in jail. Prostitution is not criminalized in these states, but living off a prostitute and coercing a minor are. So there are legal mechanisms that can be used to target those who exploit child prostitutes. While some argue that the legislation is incomplete, the main problem seems to be enforcement. In addition, where law-enforcement agencies are doing their best to protect minors in the sex trade, they are often badly under-resourced. Oddly, in a region where the cultural norm is that sex is only permissible within marriage, there appears to be a slight mismatch between criminal and civil law, with marriage generally permitted only at seventeen or eighteen except in extenuating circumstances. The trade in underage prostitutes is also the reverse side of a process of liberalization of some aspects of personal life, leading to what some observers see as a crisis in traditional ethical standards in the face of the worst laissez-faire attitudes imported from the West.

Central Asia is a region where family ties are traditionally strong, and both society and regional governments take a dim view of prostitution—

all the more so when minors are involved. So how is it that eleven- and twelve-year-old girls are ending up on the street? Part of the reason is economic—all these countries experienced major downturns after the Soviet Union broke up, and unemployment became a major problem for the first time. Simultaneously, the state-funded services and benefits that provided a basic safety-net for vulnerable parts of the community were badly eroded by the collapse of government revenues. Children from orphanages are at high risk of going into prostitution. With few opportunities, they become easy prey for pimps.

One of the most worrying aspects of child prostitution in Central Asia is the sale of children by their own parents. There are three gradations. In the first, a girl may be given away to some local man, even a relative, for a fee. This phenomenon is associated with extreme poverty in rural regions. In some cases, the girl becomes a second or third wife. There is a tradition of polygamy in the region, although it is banned by law, but in these cases—because of the coercion of minors—the practice is closer to slavery than marriage. Some girls are dispatched abroad, mainly to the Gulf States, for prostitution. The government of Uzbekistan, in what may be an implicit recognition that there is a problem—recently pledged to sign an optional protocol to the Convention on the Rights of the Child, covering the trafficking and prostitution of children, and pedophile pornography. Enforcing the laws as they stand remains a challenge for the region's police forces. Police lack the resources to deal with the problem—and alarmingly, a proportion of them are involved in taking kickbacks.

While some children toil out of necessity for their families, in some countries the use of child labor is a state policy. As World Day against Child Labor is marked as part of continuing efforts to stamp out the practice around the globe, there are still hundreds of thousands of underage children in Central Asia skipping school to work as unskilled laborers in cities or on farms. Children, some of them as young as seven years old, can be found working at virtually every bazaar in Central Asia. They sell anything from food to clothing and cosmetics, and preteen boys often push carts in outdoor markets while young girls from the countryside offer to work as housekeepers. The money they earn is often a lifeline for their families.[28]

Officials in Central Asia have long denied that children are forced to work, contending that the kids are helping their parents after school and that it is rural residents themselves who send their children into the fields to earn much-needed cash. In many rural areas, particularly in places like Uzbekistan, it is the government that forces children to pick cotton.

The practice has been in existence since the Soviet era and continued till the Central Asian countries gained independence in 1991—even after they joined international agreements banning child labor. Human rights activists say that cotton brings cash to the state coffers as well as to the pockets of the ruling elite in almost all Central Asian countries. The use of children in cotton picking has become a national duty. About 200,000 Tajik children are forced to do hard and harmful work with the number increasing during the main harvest season. In Uzbekistan, the use of child labor in the cotton sector is a state policy. As the cotton harvest begins in September, schools are shut down and thousands of children are bused to fields, sometimes with a police escort. They pick what is dubbed the "white gold" that brings around $1 billion in annual exports for Uzbekistan. In Turkmenistan as well, child labor is widely used during the cotton harvest, although the country is a signatory to the UN Convention on the Rights of the Child. It also passed laws in 2002 and 2005 prohibiting the employment of children under the age of sixteen and regulating a child's right to protection from exploitation. The late Turkmen president Saparmurat Niyazov frequently issued statements on the necessity of ending child labor, but the situation remained largely unchanged throughout his presidency. In Kazakhstan, children work in cotton and tobacco fields and as unskilled laborers in urban areas. In recent years, children from neighboring Uzbekistan and Kyrgyzstan have been working in Kazakhstan along with their parents.

## THE SOCIAL SITUATION IN UZBEKISTAN

Uzbekistan in 1991 emerged as a sovereign country after more than a century of Russian rule—first as part of the Russian empire and then as a component of the Soviet Union. Positioned on the ancient Great Silk Road between Europe and Asia, majestic cities such as Bukhara and Samarkand, famed for their architectural opulence, once flourished as trade and cultural centers. The country's political system is highly authoritarian, and its human rights record widely decried.

Uzbekistan is the most populous Central Asian country and has the largest armed forces. There is no real internal opposition and the media are tightly controlled by the state. The country is one of the world's biggest producers of cotton and is rich in natural resources, including oil, natural gas, and gold. However economic reforms have been slow and poverty and unemployment are widespread.

Two-thirds of the population of Uzbekistan lives in the countryside. At the beginning of 2002, the population of Uzbekistan reached approximately 25.1 million inhabitants, which amounts to about 40 percent of the total population of Central Asian countries. This is largely the result of the previously high fertility rate, and subsequent population growth. For many years, the government of Uzbekistan encouraged women to have many children in order to increase the population size of the country. Those who had seven or more children were traditionally glorified as "mothers-heroines." They were also the recipients of a number of benefits offered by the state, such as financial bonuses, housing allowances, extensive paid maternity leave, child allowances, and so on. The total fertility rate (TFR), however, dropped from 5.7 in 1970 to 2.4 per person in 2002.[29]

Rapid population growth produces a young population, which implies a higher burden on the society, as only a small proportion of its working population is enrolled in the labor force and produces goods and services in order to support the nonworking segments of the population. The available data reveals that about 38.7 percent of the total population of the country is under sixteen years of age, declining from 43.1 percent in 1991. The highest number of children and teenagers reside in rural areas (67.6 percent), particularly between the ages of eight and fifteen(72.4 percent).[30]

Many children live in regions with high rates of both population density and unemployment, such as Fergana, Kashkadarya, and Samarkand. Furthermore, it is important to mention that, in some regions, such as Kashkadarya, Surkhandarya, Djizak, and Samarkand, there are higher than average numbers of children. Tashkent has the lowest number of children, accounting only for 27.5 percent of the total number of inhabitants. It is evident that in Uzbekistan most children live in the countryside, which is due to the national tradition of having large families and for children to support parents in their old age. In rural areas, children under sixteen years old make up 41.8 percent of the total number of the population, against 33.1 percent in urban areas.[31]

Demographic, social and economic factors create objective preconditions for the use of child labor in the countryside. In rural areas, 31 percent of families have six to seven members, and 11 percent of families have nine and more members. Currently, 136 children are born per 100 women of reproductive age, with 152 children in rural areas.[32] During the transition to market economy, it has been observed that the number of children has decreased in both urban and rural areas, as the political and economic changes had demographic implications. Moreover, approximately 27 percent of the population lives below the poverty line and thus cannot satisfy their basic needs. In rural areas, approximately 35 percent have a higher chance of becoming poor, and 58 percent—the poorest.

According to the Constitution of the Republic of Uzbekistan, the state guarantees to all citizens the right to free education. The general secondary education is compulsory and includes grades I–IX. In addition, the law "On Education" guarantees equal rights to education irrespective of sex, language, age, race, nationality, religion, social origin, occupation, and duration of residence in Uzbekistan (Article 41). Nevertheless, in the post-Soviet period, in the more impoverished regions, the school attendance rates have decreased.[33] Many poor households cannot afford textbooks, clothes, and other items. Children have to miss school and to support their parents financially.

According to the law "On Employment of the Population," the unemployed are defined as able-bodied persons over sixteen years of age, who have no employment or income, are registered at the local labor assistance office as seeking employment, and are prepared to work and/or to enroll in vocational training and retraining courses and to improve their professional skills. According to official data, the unemployment rate in Uzbekistan is 0.4–0.5 percent (the lowest rate among the Commonwealth of Independent States), although this indicator does not reflect the real situation in the labor market. The results of an analysis of family budgets, carried out by the World Bank, using ILO methodology, placed the real rate of unemployment in Uzbekistan at 5.8 percent (in urban areas, 5.3 percent; in rural ones, 6.2 percent).[34] In Uzbekistan, unemployment affects mostly the youth, as more than 50 percent of the population are young people between sixteen and twenty-five. About 13 percent of young people in this age group are unemployed in both rural and urban areas.[35] They are generally graduates of high school, technical schools, colleges, and other educational institutions.

Today, Uzbek children work side by side with adults. They carry out the most unskilled work, such as cleaning houses and offices, washing cars, hauling goods in markets, helping parents or relatives in retail trade, and working in the agricultural sector. There are no exhaustive and reliable statistics on the number of working children. Nearly 10 percent of children between the ages of fourteen and seventeen—the age group most likely to be involved in the "black" labor market—are not included in the official statistical data.[36] Children between ten and fifteen years of age are employed for all types of work, whereas children from the five-to-nine age group do the less skilled jobs.[37] The high level of employment demonstrates that children between ten and fifteen years of age perform work suitable for and on a par with adults, as vendors, waiters, and such. One of the main reasons of participation of children in labor markets is their wish to earn money and thus assist their parents. Children have become a cheap and dependable source of labor in the country's shadow economy.

During the Soviet period, the cotton industry was central to the country's economy and the state frequently resorted to the labor of schoolchildren due to time constraints associated with cotton picking. When the country became independent, the state policy on agriculture was reassessed—the number of cotton fields were reduced and producers became more interested in independent entrepreneurial activities. Nevertheless, the use of child labor in cotton fields continued. As school authorities do not have any effective instruments to force children to go to cotton picking, they resort to softer methods of persuasion. Cotton picking is paid work. In 2001, the payment for one kilogram of picked cotton ranged from 22 somsat the beginning of the season to 15 somsat the end of the season. Working hours depend on the age of children. Pupils pick cotton between September and the end of November, as schools are closed during the cotton-picking season.[38] Recently, governmental procurement quotas were introduced for some agricultural crops that have strategic significance for the national economy. Rural students play the role of the labor reserve that ensures that the government quotas are reached. Presently, children are employed in cotton picking in an organized manner only after the best part of the cotton has been picked and when work in cotton fields does not yield substantial earnings for agricultural producers. At the beginning of the cotton picking season, agricultural producers do not allow other people to work in the cotton fields.

## CHILD LABOR STUDIES IN UZBEKISTAN

In the modern period, with the advent of a new industrial culture, the practice of child labor has assumed an alarming proportion and attracted the attention of social workers, philanthropists, and social activists to its various negative aspects. It was never a problem until the need for more production at cheaper costs, and the motive to earn more and more resulted in the use of child labor. Various social and human rights organizations took serious notice of this practice, which attracted the attention of the government, parliamentarians, and social scientists. There are some scholars, NGOs, and social scientists who have made an attempt to analyze this problem. Our endeavor will be to draw inferences from their studies and see how they can help us in understanding the problem.

The study *Invisible to the World? The Dynamics of Forced Child Labor in the Cotton Sector of Uzbekistan* by Deniz Kandiyoti has provided an evidence-based assessment of the recruitment of school-age children for the cotton harvest in Uzbekistan.[39] It draws on the results of a survey administered in six rural districts in 2007 that examines the scale and mechanisms of recruitment, the conditions of work and the consequences for children's welfare. The findings point to widespread compulsory recruitment of children for the cotton harvest for up to two months, which results in interruptions in schooling as well as exposure to conditions of work. Child labor in Uzbekistan is symptomatic of a systemic failure to establish incentives that could stimulate the growth of agricultural incomes and initiatives to reduce dependence on cotton exports. Nonwritten directives to recruit children are conveyed by local authorities to schools and local farmers. Farmers are charged with paying harvesting wages and providing transportation and other amenities. The children's wages are received by school administrations and teachers who distribute the pay weekly. The research divulges that in order to assess the scale of child labor in Uzbekistan, major international institutions with a mandate to safeguard child and labor rights have mainly relied on instruments such as the Multiple Indicator Cluster Surveys (MICS), conducted with technical support from UNICEF and UNFPA. Seasonality effects (due to surveys carried out in summer and at the end of winter instead of during the harvest season) have resulted in significant biases, leading to an underestimation of the scale of child labor. International NGOs have relied on smaller-scale but more reliable studies. Although Uzbekistan has signed two ILO conventions against child labor and two ILO conventions on forced labor and is also a signatory of the Convention of the Rights of the Child, information from throughout the cotton growing regions shows that the government continues to promote the mass mobilization of children to harvest cotton. While the government of Uzbekistan refused to allow the ILO or any other credible international observer access to the country, reports from local human rights activists indicate that schoolchildren and college students were forced to work in the cotton fields for more than two months, missing school from September until the beginning of December.

Due to continued international attention to, and criticism of, Uzbekistan's practice of forced child labor, the authorities have sought to minimize the publicly visible evidence of their involvement in the practice. Thus they stopped overseeing the safe transport of children to and from the cotton fields. Unless they brought drinking water from home, children were forced to drink unhealthy water from canals and ditches. They ate their food sitting on the ground beside the cotton fields, where pesticides and herbicides are widely used. There were no medical personnel attending to their health needs and the physicians themselves have also been mobilized to pick cotton. While in many developing countries child labor is driven by poverty, in Uzbekistan the situation is entirely different.

Uzbekistan is the world's sixth-largest producer of cotton, and the third-largest exporter according to the study *We Live Subject to Their Orders*, a three-province survey of forced child labor in Uzbekistan's 2008 cotton harvest.[40] For decades, it has used the forced labor of its schoolchildren starting in the early primary grades, college and university students, and civil servants, to harvest that cotton by hand. Unlike child labor in agricultural sectors in some other countries, this practice is organized and controlled by the central government. Each fall, shortly after the start of the school year, the government orders schools to close and school administrators to send the children out to the fields, where they remain until the cotton harvest is brought in. The report is based on seventy-two interviews in three different provinces with participants in the fall 2008 harvest.

The study *Uzbekistan: Forced Child Labor in Uzbekistan's 2008 Spring Agricultural Season* purports to offer proof that Tashkent, despite ratifying two international agreements designed to discourage the use of child labor, is continuing to send school-age children into the fields.[41] "Children suffered heatstroke, burns, and a variety of infectious diseases from poor working conditions. School hours were truncated. And for some periods schools were closed altogether to spur children into the fields. The government has increased pressure on those it suspects of transmitting any news regarding child labor." Parents who tried to keep their children in school and out of the fields were subjected to official pressure, the report said. A favorite tactic, according to the report, was public humiliation during meetings of neighborhood committees, or *Mahallas*. Those families that fail to send their children to pick cotton are criticized.

Another report, *Child Labor and Cotton Picking in Uzbekistan*, by Erkin Ahmadov , discusses t he issue of using child labor for picking cotton, which has been on the table for a long time in Uzbekistan.[42] However, the existence of forced child labor as such was strongly denied by the Uzbek authorities, claiming that "children work in the agricultural industry on a legitimate and secure basis, as all other kids in the world." In light of all the issues that child labor brings into discussion, it is remarkable to note that the practice as such was common and acceptable for many years. In spite of internal discontent and calls for prohibition by the international community, very little has been done to abolish or eradicate child labor. Now that important economic interests are at stake, it seems that the children of Uzbekistan are provided with a legal basis to protect their rights. However, in a country that gets its major sources of income from cotton exports and desperately needs "free hands" to do the work, it seems to be quite difficult to secure the rights and freedom of the most vulnerable, as their economic conditions do not allow them to leave aside even such a hard-earned and small income.

*Forced Labor in Uzbekistan's Cotton Industry*, by Anti-Slavery International, discusses the state-sponsored forced labor that underpins Uzbekistan's cotton industry.[43] In the absence of mechanized harvesters, around 90 percent of Uzbek cotton is harvested by hand. Public employees, members of the public, together with children and students are forced to work alongside paid farm workers to ensure that state cotton quotas can be met. Regional government officials mobilize children as free or cheap labor during the cotton harvest in an attempt to ensure that state cotton quotas are met. Children are also used to manually weed the fields and tend the cotton plants during the growing season. Children are also compelled to apply pesticides to the growing crop. Children who run away from the cotton fields, or who refuse to take part, can face expulsion from school. After weeks of arduous work and poor accommodation children can be left exhausted and suffering from ill-health and malnutrition. Children receive little or no pay for their labor. Some are not paid anything once deductions for food, supplies, and transport are made. Parents note that payment often falls far below the costs of replacing clothes damaged while picking cotton.

It is clear that state-sponsored forced adult and child labor remains a profound and widespread problem in Uzbekistan, despite government denials and action taken in recent months. There is a vast disparity between legal commitments made to eradicate forced labor and their practical implementation, as witnessed by the continued use of forced labor in the most recent autumn 2008 cotton harvest. Journalists and human rights defenders exposing the issue have been subject to harassment and arrest. Independent monitoring of the harvest has been very limited, and only undertaken at personal risk to journalists and human rights defenders. About the Soviet legacy the study delineates that the cotton monoculture in Uzbekistan at the expense of all the other crops is a legacy inherited from the former Soviet Union. The Soviet regime had designated certain of its republics as highly specialized producers of certain commodities. Uzbekistan was designated as a major producer of raw cotton. In order to expand the production of cotton in Uzbekistan, the Soviet Administration increased its planted acreage with massive irrigation and intensive use of fertilizers. This led to the forced labor of millions of people on state-owned farms. In the 1930s, thousands of special settlers (Karachays, Crimean Tatars, Meskhetian Turks, Russian Germans, and many others) were deported to Central Asia and were forced to cultivate cotton. By the end of the 1930s the USSR had become self-sufficient in cotton. The mass mobilization of children was one of the characteristics of cotton production during the Soviet regime. Rural schoolchildren from the ninth grade (fourteen years old) and above were forced to pick cotton up to two months every year. Public employees, including doctors, teachers, accountants, and many others were also mobilized and forced to harvest cotton. After Uzbekistan achieved independence from the former Soviet Union, the situation deteriorated significantly. Now the mobilization involves children at a younger age than ever before, due to significant drop in the level of mechanization in cotton harvesting, which has decreased from 50 percent prior to independence to just 10 percent now.

Shukhrat Ganiev and Sukhrobjon Ismoilov in their study *Child Labor and Cotton Picking* reveal that cotton remains a major agricultural crop in Uzbekistan and, most rural communities, including the children within them, have become pawns in government policies that stress high yields of cotton to constitute huge gains in the national budget.[44] Forty-three percent of Uzbek cotton is exported to Asia and 19 percent to Europe. The government artificially forces prices to be as cheap as possible—employing approximately four hundred fifty thousand children all over the country to keep costs down. Local governments and educational institutions help the central government push children at secondary schools, colleges, lyceums, institutes and universities to cotton pick. Cotton production in Uzbekistan is based on exploitation of the population. Millions of poverty-stricken rural residents work in cotton fields for scanty payment or even free of charge. Forced child labor and other types of abuses are considered commonplace. High profits then get divvied up between small groups of state elites with powerful political connections. Such an economic system is viable only under political repression, normally triggering the mobilization of mass labor working below market prices. The political regime in Uzbekistan is regarded as one of the most repressive in the world. No democratic elections are held. Opposition activists and human rights defenders are persecuted. Lack of public mass media lets most abuses go unexposed. All levels of the governments, including local governments, are involved in these abuses because accountability and responsible governance do not exist. The cotton magnates in Uzbekistan are interested in maintaining this corrupt and non-democratic political regime.

The cotton industry revolves around using the cheapest labor possible, and in most cases, this means employing children. In Uzbekistan, school-aged children are regularly involved in forced cotton picking. They spend up to two months out of the school year in cotton fields. Educational institutions and local governments condone students missing classes in order to participate in cotton picking. Living and work conditions for children involved in cotton picking are often cruel and inhumane, with most suffering degrading treatment at the hands of their overseers. Each year children involved in forced cotton picking become ill or die.

## NOTES

1. L. B. Punecha, *Child Labour: A Social Evil* (New Delhi: Alfa Publications, 2006), 2, 5, 33.
2. Kailash Satyarthi and Bupinder Zutshi, *Globalisation, Development and Child Rights* (New Delhi: Shipra Publications, 2006), 8, 11, 12, 29, 30.
3. Bilal Bhat, *Rehabilitation of Child Labour: Problems and Prospects* (New Delhi: Shipra Publications, 2011), 14.
4. Joseph Gathia, "Approaches to Combating Child Labour in India," *Caritas India Quarterly* 48, no. 4 (1998): 12.
5. UNICEF, *The State of the World's Children 2004* (New York: UNICEF, 2004), 7.
6. UNICEF, *First Call for Children: World Declaration and Plan of Action from the World Summit for Children* (New York: UNICEF, 1990).
7. Ibid., 6.
8. P. Anandharajakumar, *Female Child Labour in India* (New Delhi: APH Publishing House); UNICEF, *First Call for Children*, 13–14.
9. "Children under the age of fourteen years may not be employed or work in any public or private agricultural undertaking, or in any branch thereof, save outside the hours fixed for school attendance, the employment shall not be such as to prejudice their attendance at school." ILO, Convention 10—Minimum Age (Agriculture) Convention (1921), Article 1, http://www.ilo.org/dyn/normlex/en/f?p=NORMLEXPUB:12100:0::NO:12100:P12100_INSTRUMENT_ID:312155:NO.
10. June 26, 1973, 1015 U.N.T.S. 297(Convention 138) entered into force June 19, 1976.
11. G.A. res. 44/25, annex, 44 U.N. GAOR Supp. (No. 49) at 167, U.N. Doc. A/44/49 (1989), entered into force September 2, 1990.
12. UNICEF, *First Call for Children: World Declaration and Plan of Action from the World Summit for Children, Convention on the Rights of the Child* (New York: UNICEF, 2000).

13. Elisha Chanlett and G. M. Morier, "Declaration of the Rights of the Child," *International Child Welfare Review* 22 (1968).

14. In 1976, the General Assembly proclaimed 1979 as the International Year of the Child. The general objectives were to promote the well-being of children, drawing attention to their special needs and encouraging national action, particularly for the least privileged and those at work. It was only during the IYC that international attention became fully focused on the problem of child labor. The IYC acted as a spur to international agencies and civil society actors alike.

15. Alfred De Savza, ed., *Children in India* (New Delhi: OXFAM India).

16. Kofi Annan, *We the Children* (New York: UNICEF, 2001), 80–82.

17. Note that "child labor" is a narrower concept than that of "working children" or "economically active children." It includes all working children five to eleven years of age; excludes those in the twelve-to-fourteen age group engaged in "light work"; and, from among the fifteen- to seventeen-year-olds, includes only those in hazardous work or other WFCL. For more details see Statistical Information and Monitoring Programme on Child Labour (SIMPOC).

18. Environmental Justice Foundation, *White Gold: The True Cost of Cotton—Uzbekistan, Cotton and the Crushing of a Nation* (London: EJF, 2005).

19. Group of Human Rights Defenders and Journalists of Uzbekistan, *Forced Child Labour in Uzbekistan's 2007 Cotton Harvest: Survey Results* (Tashkent: Author, 2008), available at: http://www.laborrights.org.

20. International Labour Rights Forum and Human Rights Defenders in Uzbekistan, *Forced Child Labour in Uzbekistan's 2008 Spring Agricultural Season* (Washington, DC: Author, 2008), available at: http://www.laborrights.org.

21. "Elderly people, breastfeeding women ordered to pick cotton in Samarkand Region," *Uznews.net*, October 16, 2008.

22. Human Rights Watch, *Human Rights Watch World Report 2000: Uzbekistan* (Geneva: HRW, 2000).

23. Bilal Bhat, "Socioeconomic Dimensions of Child Labour in Central Asia: A Case Study of the Cotton Industry in Uzbekistan," *Problems of Economic Transition* 54, no. 1 (May 2011): 84–99.

24. Ibid.

25. Ibid.

26. International Crisis Group, *The Curse of Cotton: Central Asia's Destructive Monoculture* (Bishkek/Brussels: ICG, February 28, 2005).

27. Institute for War and Peace Reporting, Lost Children of Central Asia: Underage Prostitution in Kazakstan, Kyrgyzstan, Tajikistan and Uzbekistan (London: IWPR, 2004).

28. Radio Free Europe, June 12, 2008, http://www.rferl.org/featuresarticle/2008.

29. State Department of Statistics, *Population of the Republic of Uzbekistan 1991–2001* (Tashkent: Author, 2002), 79.

30. Ibid., 35–43.

31. State Statistical Agency, *Uzbekistan in Numbers 2002—Statistical Overview* (Tashkent: Author, 2003), 35.

32. A. V. Gonorskya, *Maintenance of the Full Employment of the Labour Force, Using the Potential of Non-governmental Organizations* (Tashkent: Author, 2002), 50.

33. UNICEF, "Social Monitoring—Analytical Material."

34. World Bank, "Labour Market and Unemployment in Uzbekistan" (2002), available at: http://siteresources.worldbank.org.

35. Tashkent Centre for Social Research, "Tahlil," in *Child Labour in Uzbekistan* (Tashkent: Author, 2002), 9.

36. Ibid.

37. Understanding Children's Work, www.ucw-project.org.

38. Tashkent Centre for Social Research, "Tahlil," 75.

39. Deniz Kandiyoti, *Invisible to the World: The Dynamics of Forced Child Labour in the Cotton Sector of Uzbekistan* (London: SOAS, University of London, 2008).

40. International Labour Rights Forum and Uzbek Human Rights Group, *We Live Subject to Their Orders: A Three-Province Survey of Forced Child Labour in Uzbekistan's 2008 Cotton Harvest* (Tashkent: ILRF, 2009).

41. International Labour Rights Forum/Human Rights Defenders in Uzbekistan, *Uzbekistan: Forced Child Labour in Uzbekistan's 2008 Spring Agricultural Season* (2008), Child Rights International Network, available at http://www.crin.org.

42. Erkin Ahmadov, *Child Labour and Cotton Picking in Uzbekistan* (Baltimore: Silk Road Studies Programme, Central Asian and Caucasus Institute, Johns Hopkins University, 2008).

43. Anti-Slavery International, *Forced Labour in Uzbekistan's Cotton Industry* (July 2009), available at http://www.antislavery.org/english/.

44. Shukhrat Ganiev and Sukhrobjon Ismoilov, *Child Labour and Cotton Picking* (Rapid Response Group-Uzbekistan, 2008).

# TWO

# Locating Child Labor in Uzbekistan

The resurgent interest in child labor has been seen in a combination of new international conventions, new actors, new investment, new research, new information, new ideas, and expanded activity in a wide variety of developing countries' economic and social settings. One result of this movement is increased diversity of both thinking and action. Discussion of child labor is now marked by vigorous debate in regards to a number of important issues. Even among recognized experts, both researchers and activists, there is wide and often heated disagreement even over such basic matters as what constitutes child labor, what causes it, what its affects are on children and society, and how best to deal with it. This diversity of viewpoints should be encouraged because the creative tension between them can, if constructively managed, generate more effective action and cooperation against abuse of children in and through their work.

A review of the "child labor" literature and experience reveals four highly generalized lines of thinking that can be thought of as: (1) a labor market perspective, (2) a human capital perspective, (3) a social responsibility perspective, and (4) a child-centered perspective. It is suggested that each of the most important positions now encountered in current national and international discussion of "child labor" tends to fit into one of these perspectives more easily than any of the others, and can be profitably understood in that context. The first two perspectives are of an economic orientation, while the latter two tend to emphasize social and cultural factors.

## THE LABOR MARKET PERSPECTIVE

This perspective, the dominant international paradigm of government child labor intervention for nearly a century, is fundamentally driven by mostly Western cultural notions of childhood as properly a work-free period and by anxiety about the potential impact of child workers on adult labor markets. While concern about workplace hazards on children's health and development has also been a factor, and has even been emphasized in political rhetoric, policies and programs generated under this conceptual framework have been designed primarily to separate children from work until at least mid-adolescence, rather than to make their workplace safer or more conducive to their development.

The "'labor market perspective'" takes a particular view of children, envisaging them as essentially innocent, ignorant of the world and incompetent to fend off its evils or even to recognize their own best interests. They are depicted as helpless victims, or potential victims, dependent on protection and rescue by adults. This is primarily a modern Western notion of childhood that is historically and anthropologically unusual not only for the radical division it draws between childhood and adulthood, but also for valuing children's helplessness rather than usefulness, and artificially extending their dependency to an advanced age by deliberately delaying instruction in certain life skills essential to survive, make one's living, or raise a family. Such a view of childhood leaves children free of responsibility and obliges adults to make decisions on behalf of the young that children, owing to their incompetence by definition, are deemed unqualified to make for themselves. There also is a built-in assumption that what is best for adults probably is good for their children as well.

## THE HUMAN CAPITAL PERSPECTIVE

This approach views the work of children through the lens of national economic development. It regards child labor as a product of economic underdevelopment, and suggests that the remedy is to eliminate poverty and its causes. At the macroeconomic level this entails raising the gross national product (GNP), and at the microeconomic level providing enhanced income options for the poor. It conceives of the child labor problem in terms of work and working conditions that undermine children's eventual contribution as adults to national economic development and their own economic progress, and makes no objection to children's work, per se. However, a "human capital perspective" would worry about work that stands in the way of children receiving an education, and for this reason economists and others working within it have produced a considerable literature looking at the relationships between children's work, education, and lifetime earnings. The interpretation of findings from literature, which appear ambiguous, is currently a matter of intense discussion. This perspective promotes policies and activities to develop in children the skills, attitudes and other capacities—the "human capital"—they need to contribute to economic development and become prosperous adults. It judges children's work according to whether it contributes to or detracts from this objective, opposing work that deprives children of education, for example, but approving apprenticeships or other work arrangements that transmit skills.

"The decision to educate one's children has an inter-temporal aspect, as discussed by many authors . . . [who] make a direct connection of human capital formation to child labor when evaluating the efficiency characteristics of household decisions."[1]

## THE SOCIAL RESPONSIBILITY PERSPECTIVE

This perspective regards the work of children in the context of social rather than economic development. It arises out of concern about social inequality, many types of discrimination, unjust concentrations and use of economic and political power, cultural alienation, dysfunctional family and community relationships, social irresponsibility, and the deterioration of values and moral fiber. The central concern is with the "exclusion" of disadvantaged groups from full participation in the protection, benefits, and opportunities of society, and the proposed remedy is greater social inclusion of those being excluded or marginalized.[2]

## THE CHILD-CENTERED PERSPECTIVE

Unlike the other conceptual frameworks discussed above, "child-centered" interventions in child work have children as their primary clientele, putting their interests first and foremost without filtering them through prior adult agenda. Unsurprisingly, this perspective is most associated with organizations for the defense of children. It is prominently represented at the international level by UNICEF and international child rights organizations such as the Save the Children Alliance and at the national level by a huge number of nongovernmental and community organizations promoting children's rights and welfare. It is driven by concern about conditions that impair children's growth and violate their rights. Accordingly, it conceives of child labor as that work which undermines children's well-being and individual and social development, and it judges the appropriateness of any work according to its effect on a child. Work is broadly defined to include much more than economic participation; for instance, girls working in their homes are included in the focus. The purpose of intervention in children's work is to guarantee their rights, welfare and development.

## THEORY OF INEQUALITY, PRODUCTIVITY, AND CHILD LABOR

A recent theoretical literature has linked reductions in income inequality to reductions in child labor in countries that are relatively well-off, but has not explored how income distribution affects child labor in very poor countries. We show that while in higher-productivity countries with child labor, a more equal income distribution will reduce or eliminate child labor, in low-productivity countries, a more equal distribution of income will exacerbate child labor. Econometric specifications studying child labor among ten- to fourteen-year-olds yield results generally consistent with these predictions. Policy actions that aim to bring about more equality so as to reduce child labor will likely not have the desired effect unless a country in which they are taken is sufficiently wealthy.

## KARL MARX

The results of capitalists buying children and underage persons are physical deterioration and moral degradation. Karl Marx provided a detailed brief for the General Council (International Workingmen's Association, January 28, 1868) on social issues. Child labor was deemed such an issue. Marx's position on child labor as "a progressive, sound and legitimate tendency, although under capital it was distorted into an abomination" met with no opposition.[3] Marx believed that no child under nine years of age should work. He then divided older children into three age groups—of nine to twelve, thirteen to fifteen, and fifteen to seventeen years—and suggested that they should be allowed to work two, three, and six hours per day, respectively. Marx was a supporter of part-time education, and was skeptical about the role of the state in education (he had Prussia in mind).[4] The Marxian perspective on child labor fits completely to the interpretation of child labor in the cotton industry of Uzbekistan. The capitalist elite of Uzbekistan utilize the labor of innocent children for their nefarious capitalist interests. The children working in the cotton sector in the Marxian perspective are the proletariats who spend their blood and sweat to fill the pockets of the bourgeoisie class of the autocratic state of Uzbekistan. By putting these feeble souls to such tedious and inhumane tasks the capitalist state machinery extracts as much labor as possible through the exploitative institutions to earn heavy margins for itself.

## THEORETICAL EXPLANATIONS OF CHILD ABUSE

A number of explanations have been given by scholars to explain the motivational factors in child abuse. Of these, the important ones are: (1) psychiatric explanation; (2) socio-cultural explanation which includes (a) social situational explanation, (b) social habitability explanation; (3) resource explanation; (4) social interactional explanation; and (5) social learning explanation.[5]

The psychiatric explanation was propounded by scholars like Steele and Pollock (1968), Kempe (1972), Gelles (1973), and Parke and Collmer (1975). It links with child abuse factors such as mental illness and personality defects or intra-individual abnormalities. It also links abusive parents' own childhood experiences to the individuals' weak personality development and poor self-control (Wolfe, 1987: 45).The thesis that personality disorder is responsible for child abuse was further advanced by reports that abusers often had a propensity for impulsive and/or anti-social acts that extended beyond the preventive role. A parent, according to this explanation, may abuse his/her child due to unmet emotional needs (that signify discontentment, anger or irritability), an inability to balance the child's needs and capabilities with own (parental) expectations, or emotional scars from their ability to care for their own offspring (Wolfe, 1987: 45).

This explanation initially drew support from many fields, including law-makers and public-interest groups because it directed most of the responsibility for abusive behaviour squarely at the individual involved, and absolved society from blame in contributing to the risk of child abuse through lack of education, adequate housing, family support programmes, employment opportunities, and so on. However, recent researches have disproved the role of psychopathology in child abuse. The socio-cultural explanation, given in the 1960s, maintained the external forces or socio-demographic variables within the society caused child abuse. This explanation includes three sub-explanations: social situational, social habitability and social control.

The social-situational explanation proposes that abuse and violence arise out of two factors: structural stress and cultural norms. As the social structure in which a parent lives becomes more stress ridden (or is perceived as more stressful), the greater becomes the possibility that family violence will surface as an attempt to gain control over irritating, tense events. Cultural sanctioning of violence as an appropriate conflict resolution technique further provides a foundation for the use of corporal punishment in child rearing.

If a parent was frequently exposed to harsh physical punishment as a child, he/she may have greater propensity towards viewing such behaviour as normative and inhibition against physical force may be lessened (Bandura, 1973). Steinmatz and Strauss (1974) have maintained that factors such as low income, unemployment, isolation, unwanted pregnancy and conflict with spouse/in-laws cause structural stresses which, combined with cultural acceptance for violence to resolve conflict, lead to the use of force and violence against children at home. . . .

The major problem in this explanation, according to Fieldman, (1982) is that it is unable to account for the finding that given the same set of deprivation or adverse conditions, many parents do and others do not abuse their children.

The social hospitability explanation was proposed by James Garbirino in 1977. According to him, the nature of child maltreatment depends upon the quality of the environment in which the person and family live, or the level of family support in the environment. The lesser the family support the greater the risk of maltreatment of children.

The social control explanation was propounded by Gelles in 1973. According to him, parents use violence against their children because they have no fear of being hit back or being arrested. . . .

Thus, violence is used when: (1) the cost of being violent is to be less than the rewards, (2) the absence of effective social control over family relations decreases the cost (of one member being violent towards another), and (3) family structures reduce social control in family relations and, therefore, reduce the costs and increase the rewards of being violent (Gelles and Comell, 1985: 121). . . .

. . . Gelles (1973) has maintained that certain types of children—like the handicapped, ugly, demanding, premature are at a greater risk of being abused by their parents.

This is because . . . when a parent perceives the cost of parenthood to outweigh the rewards, he uses violence against his children. Ivan Nye (1979) like Gelles, had also earlier accepted the application of Peter Blau's theory, in explaining child abuse.

He has proposed that child beating is less common in families that have relatives and/or friends nearby. . . .

This explanation has been criticized on the ground that: (1) It is preposterous to assume that relations between parents and children are based on reciprocity and that parents' treatment of children is determined in terms of calculations of rewards and costs. (2) Assuming it is so, why do all the parents not make such calculations and only a few do it, that is, why do all the parents not beat their children and only some parents indulge in child beating? . . .

The resource explanation was given by William Goode in 1971. According to it, the use of force by an individual depends upon the extent to which he can command or master the resources—social, personal and economic. The more resources a person has, the less he/she will use force in an open manner. Thus, a father who wants to be a dominant person in the family but has little education, low prestige job, low income and lacks inter-personal skills may choose to use violence against his children to maintain the dominant position.

The social interactional explanation was given by Burgess in 1979. It approaches the etiology of child abuse in terms of the interplay between individual family and social factors in relation to both past (for example, exposure to abuse as a child) and present (for example, a

> demanding child) events. The parents' learning history, inter-personal experiences and intrinsic capabilities are regarded as predisposing characteristics presumed to be important contributors to an abusive pattern. In this explanation, the potential role of a child in provoking abuse is also acknowledged. . . .
>
> . . . The interactional explanation, thus, is not necessarily limited to observable behaviour alone (like parental criticisms or displays of anger . . .) but includes cognitive and effective processes too (like intelligence, attitudes . . .) that may mediate behavioural changes.

## THEORETICAL MODELS/PERSPECTIVES ON SOCIAL PROBLEMS

Every science, including social sciences like sociology, accumulates knowledge through interplay between theory and research. A theory is a set of statements that explains the relationship between phenomena. The key role of theories is to tell us why something occurred. They help us organize the data from research into a meaningful whole.[6]

As theory provides a point of view for the study of specific problems, an attempt is made here to introduce some of the relevant theories with the ultimate aim of applying them in the empirical data analysis and discussion.[7] Major theories covered are:

1. structural functional theory
2. conflict theory
3. production relations
4. exchange and reference group theory

### *Structural Functional Theory*

Structural functionalism has its family tree in the work of early sociologists, especially Emile Durkheim and Max Weber. Among modern-day scholars, it is most closely associated with the work of Talcot Parsons and Robert Merton.

The term "structure" and "function" refer to two separate but closely related concepts. Structure can be compared to the organs or parts of the body of an animal, and functions can be compared with the purposes of these structures. Like a biological structure, a social system is composed of many interrelated and interdependent parts or structures with specific functions. Central to an understanding of social structures is the concepts of "status" and "role." Simply defined, a status is a socially defined position. Status is both ascribed and achieved.

Each social system performs certain functions that make it possible for society and the people who comprise that society to exist. Each serves a purpose that leads to the continuation or firmness of the larger society. The functionalist perspective assumes that these social systems have an underlying tendency to be in symmetry or balance; and a system that fails to fulfil its functions will result in an imbalance or disequilibrium. In acute cases, the entire system can breakdown when a change or failure in any one part of the system frequently provokes changes elsewhere in the system.[8]

According to Merton, a social system can have both manifest functions and latent functions. Manifest functions are intentional and recognized; latent functions are neither intentional nor recognized. Merton recognized that not all consequences of systems are functional—that is, they do not all lead to the maintenance of the system. Some lead to unsteadiness or the collapse of a system. These consequences he termed *dysfunctions*. Robert Merton's theory of functional alternatives provides one way to avoid such dysfunctions. Functional alternatives are other ways to achieve the intentional goal.

### *Conflict Theory and Its Application*

The conflict perspective in modern sociology derives from the work of Karl Marx, who saw the struggle between social classes as the "engine" of history and the main source of social change. Among its latest proponents are such people as C. Wright Mills, Lewis Coser, and Ralf Dahrendorf. They share the view that society is best understood and analyzed in terms of conflict.

Karl Marx began with a very simple assumption: the structure of society is determined by economic organization, predominantly the ownership of property. Inherent in any economic system that supports inequality are forces that generate revolutionary class conflict.

The browbeaten classes sooner or later recognize their submissive and inferior status and revolt against the dominant class of property owners and employers. In conflict theory, it is assumed that societies are in a continuous state of change, in which conflict is a permanent trait. "Conflict" does not necessarily imply utter violence; it includes tension, competition, and disagreement over goals and values.

Conflict theory tells us to look for the hidden strains and frustrations, mainly between those in power and those under their control. Conflicting relationships are neither always explicit nor always expressed by individuals. Employees might not openly express their discontent because they feel that they are powerless to change the situation. In fact, many sociologists have suggested that people fail to express dissatisfaction not because they feel powerless to change things, but because they may not be aware that things could be better or because they are resigned to the situation.

Like structural functionalism, conflict theory is used by sociologists to (1) Explain the relationship between the parts of a social system and the inequalities that exist between these parts; (2) Discover and explain the sources of the conflict; and (3) Help create techniques to deal with conflict.

### *Production Relations*

The social relations that people enter into in the course of producing the material wealth of a society are fundamental, both in determining the structure of society and its development. The relations of production are the relations not between the human produce and nature, but between the human beings themselves as they occupy different places in society as a result of the division of labor, which is the basic character of human population.

According to Marx, the relations of production form the economic structure of society. And this economic structure is the basis not only for the other aspects of economic life, such as distribution and consumption, but also for the whole of the society with its other structures. Marx even identified the relations of the production with the relations of property as their legal expression.[9]

In Marxist theory, the mode of production can be defined as the labor process—forces of production, the relationship between the workers and the owners of the means of production, and the relationship between the workers and the product. Secondly, the mode of production is itself reproduced through the interactive process of economic, ideological, and political mechanisms intrinsic to itself, and through its subordination of or by other modes of production. Thirdly, the mode of production embodies contradictions at each of these points that both drive it forward and develop the conditions for its own disintegration through the development of class struggle and class alliance involving those enmeshed in the surrounding modes of production.[10]

Relations of exploitation or, to be more precise, the relations of surplus appropriation, are central to relations of production. This is the relation through which the immediate producers are alienated from the control of surplus of their labor.[11]

### *Exchange and Reference Group Theory*

Exchange Theory is based on the belief that life is a series of exchanges involving rewards and costs. In economic exchanges, people exchange money, goods, and services, hoping to profit or at least break even in the exchange. Exchange theory assumes that people seek rewarding statuses, relationships, and experiences and try to avoid costs, pain and punishments. Given a set of alternatives, individuals choose those from which they expect the most profit, rewards, or satisfaction and avoid those that are not profitable, rewarding, or satisfying. When the costs exceed the rewards, people are likely to feel angry and dissatisfied. When the rewards exceed the costs, they are likely to feel that they got a good deal. Both parties are more likely to be satisfied with the interaction if there is perceived equity in the exchange, a feeling on the part of both the rewards were worth the costs.

There are two different schools of thought in the exchange theory perspective. George Homans believed that behavior could be explained in terms of rewards and punishments. In exchange theory, the rewards and punishments are the behavior of other people, and thus involvement in exchanges assumes that rewards will be proportional to costs. Peter Blau is the advocate of a different school of exchange theory. Blau argued that the exchange is more subjective and interpretative and that the exchanges occur on the symbolic level. As a result, money may be a just reward only if it is defined by the receiver as such, and psychic rewards of satisfaction with doing a good job or of pleasing someone may be as important as money, gifts, or outward responses of praise. Both agreed that all exchange involves a mutually held expectation that reciprocation will occur. If resources or exchange criteria are unequal, one person is at a distant disadvantage and the other has the control and power over the relationship. If there is indeed an imbalance of rewards and costs to one or all of the parties, then there must be some form of negotiation to restore the necessary balance of exchange.

Organizations engaged in exchanges are governed by their relative resources and by the norms of reciprocity and fairness. The society as a whole becomes stratified by the same kind of differentiation that characterizes its individuals. Overall solidarity is generated by "generalized media" of exchange: norms or laws that codify the principles of exchange into abstract principles. Individuals learn these norms when socialized into society's system of common values. Possessions of these norms then make it possible for exchanges to take place at a distance, instead of through direct interaction.[12]

Reference groups, according to George Herbert Mead and Robert Merton, serve two kinds of functions. (1) They set standards for the behaviors of the individual. The individual takes the standards and value judgments of the reference groups and shapes his attitudes and behavior in accordance with them. Such reference groups are called "normative" reference groups. (2) They provide a frame of comparison relative to which the individual evaluates himself and others. They provide a context for evaluating the relative position of the individual and others. Such groups are known as "comparative" reference groups.

### *Using the Theoretical Perspectives*

The major elements of each perspective and its view on social problems have been outlined in the following table. The perspectives should be seen as three different "tools," each of which is useful in analyzing particular social problems.[13]

## CONSTRUCTING SOCIAL PROBLEMS

Use of the theoretical perspectives can be illustrated by looking at an important element of the study of social problems: how a social condition becomes a social problem. The existence of a condition alone, even when the condition produces negative consequences, does not make it a social problem; to become a social problem, it must be so defined by some group). This process of social definition or construction involves a number of elements: how and why groups identify conditions as problems, how the groups develop an understanding of the causes of the problems, and how situations are developed and implemented. Each of the three perspectives contributes to our understanding of this process.[14]

### Table 2.1

**Table 2.1. An Outline of the Sociological Perspectives**

| | Functionalism | Conflict Theory | Interactionism |
|---|---|---|---|
| **View of Society** | A system of interrelated and interdependent parts | Made up of groups struggling with one another over scarce resources | Individuals in face-to-face interaction create social consensus |
| **View of the Individual** | People are shaped by society to perform important functions for society | People are shaped by the position of their groups in society | People are symbol manipulators who create their social world through social interaction and consensus |
| **View of Social Change** | The social system tends to resist change as disruptive | Change is inevitable and continuous | Change occurs when there is no shared consensus about expected behavior and newly found consensus develops |
| **View of Social Problems** | Caused by dysfunctional activities or disorganization in the social system | Arise when group believes its interests are not being served and works to overcome perceived disadvantage | Arise when a condition is defined as stigmatizing or disruptive of normal social expectations |
| **Key Concepts** | Integration, interdependence, stability, equilibrium | Interest, power, dominance, conflict, coercion | Interpretation, consensus, shared expectations, specially created reality |

From the *functionalist perspective*, the social construction of social problems depends, at least in part, on the extent of social disruption or social disorganization produced by a social condition. Conditions that are more disruptive are more likely to be defined as social problems by significant groups or large numbers of people. But many social conditions disrupt the lives of only some people, and these conditions may be defined as problems by some groups but not others.

The *conflict perspective* helps us recognize that elites and others with access to resources or power play a greater role in this process of social definition: it is the conditions that negatively affect their values and way of life that are most likely to be defined as social problems. So, definitions of social problems are constructed out of the clash of competing interest groups.

The *interactionist perspective* recognizes the importance of symbols and social meanings in shaping human life, and it points out that defining a "condition" as a "problem" is a matter of attaching certain negative meanings to the condition. The process of interpretation is central to human social life, and people have to interpret a set of objective conditions as something that is "bad" or "negative" before they will act on it.

This brief illustration of constructing social problems demonstrates how using the three perspectives can provide a more complete understanding of a topic. This social construction process is complex and continuous, and the resulting social definitions are constantly shifting and changing.

## CHILD LABOR IN UZBEKISTAN

Each of sociology's major theoretical paradigms addresses the significance of children in social organization. Sociologists have explored questions of child labor from several theoretical perspectives, including structural functionalism and conflict theory. After going through the literature and the empirical evidence from the field in Uzbekistan, it appears the structural functional and conflict perspectives fit into the understanding and application of child labor in the cotton industry of Uzbekistan. The literature that has been generated on child labor in different sectors of the economy around the world and Uzbekistan in particular suggests the relevance of major theoretical approaches. Since child labor in the cotton industry of Uzbekistan is unique and different in its socioeconomic profile and dimension, the relevance of a particular theory to the application of the problem is extremely significant, and the whole discourse is to be seen through the prism of that perspective.

Since independence, Uzbekistan has pursued policies in support of rapid industrialization based on import substitution. On a net basis, resources have been channeled out of agriculture, which remains the key sector of the Uzbek economy, in order to finance an ambitious industrial investment program and to subsidize consumers, especially in urban areas. Resource extraction from cotton—Uzbekistan's most important crop and foreign exchange earner—has played a prominent role in this strategy. The authorities have taxed the cotton and wheat subsectors through a variety of implicit and explicit mechanisms, including the foreign exchange system, the state order system, low producer prices, and controls over agricultural collectives as well as marketing and processing organizations. All land has remained under government ownership, farm restructuring and privatization have been limited, and property rights have remained insecure. "*Milking the cow*" is the expression a government official used to describe the government's policies toward agriculture. As regards cotton and wheat production, policies continue to resemble those of the Soviet Union. Going through the operation and structure of agriculture and its processes, including the forced labor of children in the cotton industry, the structural functional theoretical framework finds relevance in understanding the whole discourse through its prism.

Although the mobilization of child labor for cotton harvests has its institutional and organizational roots in the Soviet command economy, its current persistence is best explained with reference to a combination of factors: a partial process of agrarian reform that continues to tie private farmers into compulsory crop-sowing and procurement quotas, a sharp decline in farm mechanization since independence, and a short harvesting season that creates labor bottlenecks at peak times. The sharp increase in seasonal or more permanent labor migration from rural areas to wealthier neighboring countries, mainly Kazakhstan and Russia, is both a response to increasingly precarious rural livelihoods and a further cause aggravating dependence on alternative sources of cheap labor.

The structure of the society in Uzbekistan has been structured by the political elite in such a way that the interests of the state and sub-state structures are well served. The problem is complicated and requires careful analysis. Among the important issues are the demolition of the old political system; the rejection of the former ideology; the establishment of new relations in the economic, political, and spiritual spheres of society; and the formation of a new political system. Understanding these processes and structural adjustments is essential in order to actually find the relevant theoretical frame among the various models of the child labor problem. Historically, under the Soviets, problems in sociology were understood within the confines of Marxist-Leninist principles, with their classical approach to almost all social phenomena. According to Marx the relations of production form the economic structure of society. And this economic structure is the basis not only for the other aspects of economic life, such as distribution and consumption, but also for the whole of the society and its other structures. Thus the structural functional perspective is relevant to the child labor problem in Uzbekistan today.

The conflict perspective is equally relevant to the structural and functional base of cotton cultivation and child labor in Uzbekistan. The public resentment of the cotton cultivation system is very high but is hidden due to the repressive character of the autocratic state. It is useful to point out here that forced sterilization is another repressive tool being used by the state to convince the international community that infant and maternal mortality as well as demographic composition are under check. Doctors are being given quotas for removing uteruses of fertile young women of child-bearing age, and punishments are stringent for failing to meet the quotas or for making the information public. Numbers of cases have been found where the young girls and women were unaware about the removal of the uterus as the whole process was done in a state of unconsciousness. The women learned of it only after a period of time and were very surprised and puzzled to discover that they could not bear children, which resulted in depression, stress, and many other ailments. Those who try to challenge the system face punishment, and large numbers of dissenters are now imprisoned and languishing in jails. So it becomes clear that there is a strong conflict between the goals of public agencies and the agencies of the state. But since the state is very powerful under such an autocratic structure, the public resentment against the system remains mostly hidden. Thus the conflict theoretical frame is quite relevant and applicable to cotton cultivation and child labor in Uzbekistan.

## NOTES

1. D. K. Brown, A. V. Deardorff, and R. M. Stern, *The Determinants of Child Labor: Theory and Evidence* (Ann Arbor: University of Michigan; School of Public Policy, 2002), 9.
2. Myron Weiner, *The Child and the State in India* (Princeton, NJ: Princeton University Press, 1991).
3. Quoted in S. Padover (ed.). 1975. *Karl Marx on Education, Women and Children* (New York: McGraw-Hill, 1975), 91.
4. Ibid., 33.
5. Babita Agarwal, *Child Labor in India* (Delhi: ABD Publishers, 2007), 110–14.
6. Thomas, J. Sullivan, *Introduction to Social Problems*, 5th ed. (Boston: Allyn and Bacon, 2009), 9.
7. P. Anandarajakumar, *Female Child Labour in India* (New Delhi: ADB Publishing House), 20–26.
8. Ibid., 20–26.
9. Bastiaan Weilenga, *Introduction to Marxism* (Bangalore: Centre for Social Action, 1991), 26–27.
10. H. C. Srivastava and M. K. Chaturvedi, *Rural Middlemen—Network of Patronage* (New Delhi: Ashish Publishing House, 1986), 113.
11. Venkatesh B. Athreya et al., *Barriers Broken: Production Relations and Agrarian Change in Tamil Nadu in India* (New Delhi: Sage Publications, 1990), 13.
12. Randall Collins, *Theoretical Sociology* (New Delhi: Rawat Publications, 1997), 338–48.
13. Sullivan, *Introduction to Social Problems*, 15–16.
14. Ibid.

# THREE

# Cotton Cultivation and Child Labor in Uzbekistan

Uzbekistan is the second-largest exporter of cotton in the world, selling over eight hundred thousand tons of cotton every year. Europe is its major buyer. But while the former Soviet Republic is at the forefront of global cotton production, its human rights and environmental record lags far behind the rest of the world. Forced child labor, human rights violations, excessive pesticide use, the draining of an ocean and severe poverty are all rife in cotton production in Uzbekistan.

Instead of using machines to harvest cotton, as is done in other major cotton-exporting countries, Uzbekistan's government uses children. Uzbekistan's cotton farmers are made to suffer too. Despite producing a crop worth over US$1billion, those forced to grow cotton receive little of the revenues generated from its sale. Official figures suggest that farmers receive around one-third of the value of their cotton. In practice many get far less. Instead, Uzbekistan's cotton exports, which represent around 60 percent of the state's hard currency export earnings, are appropriated by the country's totalitarian dictatorship led by President Islam Karimov.

Side by side with the human rights violations caused by cotton, is an environmental catastrophe of astonishing proportions. In order to irrigate its 1.47 million hectares under cotton, Uzbekistan's regime has all but eradicated the Aral Sea, which was already in bad shapes during Soviet times. Once the world's fourth-largest body of water, the Aral is now reduced to just 15 percent of its former volume. Appalling mismanagement of this vital water resource has witnessed the disappearance of the sea's twenty-four species of native fish from its waters, the drying out of associated wetlands, and the creation of tens of thousands of environmental refugees; the former dependents of the Aral's ecosystem.

## Table 3.1

**Table 3.1. Cotton Production in Central Asia (Thousands of Tons) and as a Share of Global Production**

| | **1913** | **1940** | **1970** | **1980** | **1990** | **1994** | **1998** | **2000** | **2002** | **2004** |
|---|---|---|---|---|---|---|---|---|---|---|
| **Kazakhstan** | 11 | 72 | 91 | 118 | 102 | 70 | 55 | 85 | 105 | 148 |
| **Kyrgyzstan** | 9 | 31 | 62 | 68 | 25 | 18 | 27 | 27 | 25 | 40 |
| **Uzbekistan** | **171** | **457** | **1483** | **2061** | **1593** | **1248** | **1000** | **975** | **1033** | **1125** |
| **Tajikistan** | 11 | 57 | 240 | 334 | 256 | 168 | 110 | 106 | 165 | 172 |
| **Turkmenistan** | 23 | 70 | 287 | 415 | 437 | 314 | 197 | 187 | 148 | 203 |
| **Total Central Asia** | **225** | **687** | **2163** | **2996** | **2413** | **1818** | **1389** | **1380** | **1476** | **1688** |
| **World production** | 6296 | 6934 | 11740 | 13831 | 18970 | 18762 | 18713 | 19437 | 19437 | 26193 |
| **% of world production** | 3.5% | 10% | 18.5% | 21.5% | 13% | 9.5% | 7.5% | 7% | 7.5% | 6.5% |

Such gross exploitation of a nation and its environment has only been possible within a framework of extreme control. President Karimov has eliminated any form of democratic representation; prohibited a free media; subverted basic civil liberties; and institutionalized the use of torture and intimidation within the police, National Security Service, and prisons. Government attitude to public protest—peaceful or not—is brutal, as most recently witnessed by the response to demonstrations in the town of Andijan in May 2005. Demonstrators were met with indiscriminate shooting, leading to an estimated seven hundred deaths and the subsequent arbitrary arrest of activists, human rights defenders, and independent journalists.

Given such conditions, the Uzbek people have been left with little option but to abide by the commands of the Karimov administration. Tellingly, those Uzbeks who have felt able to speak out are clear in their condemnation of the cotton industry and united in their view that under the current regime it does little if anything to benefit the people, but much to support a corrupt and brutal government. Despite these well-known abuses, Europe remains the major destination for Uzbekistan's cotton exports. Traders continue to associate with the regime, buying cotton in exchange for a substantial hard currency income, and high street fashion outlets sell clothes manufactured from Uzbek cotton.

## TRADE IN UZBEK COTTON

Uzbekistan is one of the "Big Five" countries that dominate global cotton production (China, the United States, India, Pakistan and Uzbekistan). However, unlike China, India, and Pakistan, which each use most of the cotton they produce; Uzbekistan has a limited capacity for domestic textile production. As a result over 75 percent of Uzbek cotton is sold on the world market every year, making Uzbekistan the second-largest cotton exporter in the world. According to data from the United Nations, the major destination for Uzbek cotton is Europe, which receives almost a third of all cotton sold by Uzbekistan. These exports have a value of around US$350 million each year.

Companies that buy Uzbek cotton are inextricably entwined in an economic system that perpetuates poverty and supports labor practices that constitute flagrant violations of human rights. But while the realities of government corruption and forced child labor are hard to ignore, the global cotton industry has done little to address the manner in which Uzbek cotton is produced. In international markets, Uzbek cotton is sold by leading European and US corporations; its production is financed by Western banks, and the final product is sold further under famous brands in Western countries. No international corporations or clothing suppliers pay enough attention to the conditions under which the cotton is produced. Moreover, no international organizations, or international financial institutions have taken enough efforts to fight against such abuses

## CHANGING TRENDS IN THE COTTON SECTOR OF UZBEKISTAN

The reliance of the Soviet command economy in Uzbekistan on "cotton campaigns" that mobilized the population, including school-age children, during harvest periods is well documented. An integrated network of institutions, from regional and local administration, to schools and collective farms were involved in securing additional labor at peak times. At first glance the use of child labor in cotton harvests, relying on an existing infrastructure of institutions, may appear as a carry-over from collective agriculture. However, the evidence points to significant changes in both the *context* and the *mechanisms* of reliance on child labor in the aftermath of agrarian reforms starting after the breakup of the Soviet Union and Uzbekistan's independence in 1991.

During the Soviet period, Uzbekistan was a region of high rural unemployment and underemployment and, compared to the rest of the Union, of low agricultural wages. These trends were aggravated further after independence when Uzbekistan's agriculture, organized around some 940 *kolkhozy* and 1,108 *sovhozy* in 1990, was gradually de-collectivized. Collective enterprises, apart from providing their members with jobs, also played an important role in allocating additional plots to households that were able to supplement their incomes by planting a second crop. The second-crop economy also absorbed some surplus labor. Privatization policies not only resulted in massive job losses but, over time, also restricted access to the second-crop economy for the majority of former collective enterprise workers.

Furthermore, after independence the terms of trade for agriculture deteriorated drastically. The government, cut off from the budgetary grant it received from the USSR, was forced to find new sources of revenue. Extraction of surplus from agriculture by driving a wedge between the procurement price and the export price of cotton was a readily available alternative. By 1994, the procurement price for cotton in real terms was a fraction of what it was in 1990. The agricultural sector continues to be subject to high levels of taxation while levels of investment in rural industries (which used to provide additional jobs during the Soviet period) have plummeted. These adverse trends have combined to deepen both unemployment and rural poverty.

## PROCESS OF AGRARIAN REFORM

Land reform is blocked in Uzbekistan. Uzbek farmers do not own the land they use; they do not even have the right to choose the crops they want to produce and the buyer they want to sell their products to. Private access to land was steadily expanded in Uzbekistan through the allocation of leaseholds to a new private farming sector that took over production from *shirkats* (collective enterprises). However, farmers continue to be tied to the state procurement system through a *shartnoma* (contract) system that specifies the particular combination of crops they are allowed to cultivate and the state delivery quotas for strategic crops, namely wheat and cotton. Producers have little control over crop allocation, access to input markets or buyers for their crops. Local *khokims* (provincial governors), who play a key role in allocating land for private farming, are still held responsible for ensuring that their region meets its procurement quotas, and they risk endangering their jobs if they fail to do so. Meanwhile, local administrators are motivated to extract as much cotton as possible from farmers, who are caught between the obligations to fulfill delivery quotas, their desire to maximize their profits, and the necessity to cut their costs as far as they can. Thus, although different players in the agricultural sector do not necessarily have identical stakes over the utilization of child labor, they share a common interest in timely access to a source of cheap labor.

A much less understood and documented effect of privatization on cropping patterns has to do with the entrance of new, more powerful players into the ranks of "new" farmers. Micro-level studies clearly suggest that their ability to farm profitably rests on their ability to negotiate activities and crop mixes that minimize their involvement in cotton farming.[1] To what extent does the ability of well-connected new private farmers to evade the unprofitable cotton crop actually increase the pressures on those who are still subject to cotton delivery quotas? Does this have a bearing on methods of labor control? Do new exit strategies of both the more privileged and of the poor increase the need for coercion in securing an adequate labor supply for cotton cultivation, including the labor of children? These issues certainly merit further, in-depth investigation.[2] It is, however, safe to conclude that the path of agricultural reform adopted by the Government of Uzbekistan is far from having created conditions that would help eradicate forced child labor and other forms of coerced labor.

## CROP CYCLE

The cotton-picking season is short and starts with the maturing of cotton bolls, usually at the beginning of September. The onset of autumn rains and cold weather reduces the quality of the cotton, which starts fetching lower prices as the harvesting season advances. The first two weeks of the harvesting season are critical. Farmers who are not able to pick the bulk of their cotton on time stand to lose financially. This creates pressure to harvest as much cotton as possible within a narrow timeframe. As the picking season advances, the quality, quantity, and pay levels of the cotton harvest decline to such an extent that there are hardly any profits to be made from this activity. But child laborers can be made to stay on the fields until the very end of the harvest period, well beyond the point when the adult labor supply has dried up.

## COTTON FARMING AND THE PATTERNS OF OUT-MIGRATION

The process of agrarian reform initiated a new period of hardship for rural populations (according to 2005 figures 64 percent of the population live in rural areas and the agricultural sector employs about 32 percent of the workforce).[3] In the initial stages of de-collectivization, *shirkats* were chronically in arrears of wages. Rural households were only able to survive through a combination of livelihood activities in the informal sector and the second crop economy. As privatization proceeded and *shirkats* were dismantled rural households started to lose their toehold in the second crop economy and many joined the ranks of a casual labor force of male and female *mardikor* (daily workers). Without the direct and indirect benefits of membership in collective enterprises, the livelihoods of rural households became increasingly precarious.

The population responded to these hardships through seasonal migration to wealthier countries with a high demand for labor, primarily Kazakhstan and Russia. Experts estimate that the total number of labor migrants (legal and illegal) from Uzbekistan to varied destinations such as Russia, Kazakhstan, South Korea, Turkey, UEA and others reach 1–1.5 million and account for up to 8 percent of the GDP in remittances.[4] The higher wages laborers are able to earn by becoming migrants (an approximate average monthly wage of US$300–500 as compared to US$200 per harvesting season in Uzbekistan) act as a magnet for able-bodied men and women who can no longer subsist in the agricultural sector of Uzbekistan.

This has increased the pressures on the operations of the cotton sector. The administration now has to combat two different types of centrifugal tendencies in order to keep up levels of production: (a) the attempts of farmers to get out of cotton production in favor of more profitable crop mixes; and (b) the attempts of laborers to exit Uzbek agricultural production altogether in favor of more lucrative jobs elsewhere.[5] This conjuncture has led to increasing levels of coercion and policing of both land use and of the agricultural labor force, pushing up demand for a cheap substitute labor force. The combination of factors detailed above points to a new set of constraints that condition the demand for child labor. If anything, the relative contribution of child and other forms of coerced labor to total output could increase unless the Government of Uzbekistan adopts a new mix of agricultural policies that can successfully break the vicious cycle of reliance on compulsion to keep up production levels. Likewise, diversification of the economy and decreasing reliance on cotton as a key export commodity could, in the longer term, assist in alternative job creation.

## FORCED CHILD LABOR AND THE COTTON INDUSTRY

In all regions of Uzbekistan, government officials mobilize children in an attempt to ensure that state cotton quotas are met. Schools are closed down, and children as young as seven are sent to the fields to pick cotton by hand. Headmasters are given quotas that dictate how much each student is to harvest. Those who fail to meet their targets, or who pick a low-quality crop, are reportedly punished with detentions and told that their grades will suffer. Children who run away from the cotton fields, or who refuse to take part, can face expulsion.

Statistics on children employed in the cotton sector in Uzbekistan are difficult to obtain, but the London-based rights group Environmental Justice Foundation (EJF) says around two hundred thousand children work in the major cotton-producing region Fergana. Fergana is one of the fertile regions in Uzbekistan and is about four hundred twenty kilometers east of the capital, Tashkent. It is impossible to estimate the exact number of children forced to pick cotton, but tens of thousands are involved each year. According to UNICEF in 2000, an estimated 22.6 percent of Uzbek children aged between five and fourteen worked at least part-time, primarily in cotton harvesting.

In Soviet times up to two-thirds of Uzbekistan's cotton was harvested by machine. Nowadays this figure has dropped to just 10 percent. Instead, the majority of Uzbek cotton is gathered by hand, often by children. According to reports from local inhabitants from nine of Uzbekistan's twelve territorial units, (Jizzakh, Fergana, Namangan, Syr Daria, Surkhandaria, Bukhara, Khorezm, Tashkent, and Samarkand provinces) by the third week of September local governments and school administrators send children as young as the seventh grade (ages thirteen to fourteen), and in some cases as young as fifth grade (eleven to twelve) out to the fields to pick cotton. By the end of September, pressure to bring in the harvest before the rains led local officials to order the lower-grade classes, from first grade on, to labor on the harvest.

In Fergana, schools are closed and children are sent out in September, though a week earlier those same schools force children to sign statements that they will remain in school after the end of the semester, possibly to give local government officials plausible deniability if the children's presence in the fields is challenged. In Namangan, every day local government officials and bureaucrats from the local education department visit the fields to check up on the number of pupils out picking, and to make sure that harvest targets are being met. Similarly the Samarkand provincial government also sent its schoolchildren out to pick cotton in September.

Most rural communities, in fact, have become pawns in government policies that stress high yields of cotton. All parts of cotton production, from planting to selling, are state controlled. Under this control, the government artificially forces prices to be as cheap as possible by employing approximately four hundred fifty thousand children all over the country to keep costs down. Local governments and educational institutions help the central government push children at secondary schools, colleges, lyceums, institutes and universities into cotton picking.

## WORKING CONDITIONS

In cotton growing areas, school officials mainly send children to assist in preparing the cotton fields and tending the plants. Work includes gathering last year's cotton bushes, plowing, planting, weeding, hoeing, and sometimes fertilizing and/or spreading pesticides. In addition, children are also sent to work on other major crops or made to gather mulberry leaves, used to feed silk worms. Others gather the silk cocoons themselves. Some children are sent to harvest spring vegetable crops, such as potatoes and onions. But majority of Uzbek children are forced to work in the cotton cultivation.

According to the law, children should not work long hours. But in practice the law is not strictly followed anywhere in Uzbekistan. The number of hours of work in the cotton sector depends upon a number of factors such as (1) demand of cotton at a particular period of time, (2) availability of manpower including child laborers, and (3) the schedule for delivery of cotton quotas. Since no study on child labor in the cotton industry in Uzbekistan has so far been undertaken in this part of the world, it is not possible to have a comparative study of the present data. However the studies so far undertaken in other parts of the world show that working hours of children are long, adequate rest intervals are missing, and adequate holidays too are absent in this hectic work. Some child laborers reach their work place early in the morning.

With a view to collecting factual information on the daily hours of work, we did a field study and interviewed child laborers at a site. The collected information reveals that the majority of them, 103 (57.22 percent), openly accepted that they were working ten to twelve hours every day. However, the United Nations Convention on the Rights of the Child prescribes five and a half hours of work for child laborers everyday and only working in special sectors of the economy not detrimental to the health and education of the children. Some of the child laborers informed us that their working hours were flexible; whenever there is demand of quotas immediately they are put to work for longer time. On the other hand, some children in cotton cultivation were working hours that were neither fixed nor flexible. Even if they worked for longer hours, they were not paid extra in spite of meeting the requirements assigned to them for the day or for the given time. Despite all these factors, eighteen (10 percent) of the children picking cotton were doing so for social and economic reasons and felt satisfied with their work conditions. But 153 (85 percent) felt exploited and had concerns for their future (see table 3.2).

Children are working for long hours in the fields during the spring season. Schools are closed for a full month before the official end of the school year in order to force the children out to work. Since the sun

**Table 3.2**

**Table 3.2. Duration of Work and Satisfaction with Work**

| S. No. | Duration of Work | No. | Percentage | Satisfaction with Work | No. | Percentage |
|---|---|---|---|---|---|---|
| 1 | 9–11 hours | 72 | 40 | Satisfactory | 18 | 10.00 |
| 2 | 11–14 hours | 103 | 57.2 | Unsatisfactory | 153 | 85.00 |
| Total | | 175 | 97.2 | | 171 | 95.00 |

comes up early and it gets hot, the children come out to the fields at 6:30 or 7 AM. They have an hour or two for lunch, and then they come out again for work until 7 or 7:30 PM. Their working day lasts nine to eleven hours. Naturally in the days when the school is closed down and lessons are cancelled, the children together with their teachers come out for the mass *khashar*. At other times during the spring season, children are sent out to the fields once the school day is complete or after a shortened school day. Schoolchildren are subject to mobilization even on the weekends, and work eleven to twelve hours per day or more. Farmers' reluctance to spend funds on gasoline requires that children must often make their way home on foot, which lengthens their work day even more. Most of the schoolchildren working in the cotton fields rarely return home before 9 or 10 PM during the spring season.

Uzbekistan's climate makes spring fieldwork particularly onerous. Cotton is grown in the irrigated steppe, or semi-desert areas, where summer temperatures can reach 45 degrees Celsius (113 Fahrenheit). By late April and early May, average daytime temperatures hover around 30 degrees Celsius (86 Fahrenheit) but can reach 35 (96 Fahrenheit). Without shade, protective gear, adequate rest periods or water, heat stroke is common. While in Soviet times it was common for farmers to provide nutritious lunches for children in the fields, it is increasingly rare now for farm administrators to arrange any meals for the children who work for them. If they do, most often children are being fed plain macaroni, or bread and tea. Usually children must bring their own food from home, which, given the low level of remuneration for this highly physically taxing work, and overwhelming rural poverty, is often a burden on families. The stark economics of hunger, inadequate food, and the lack of clean water and sanitation pose a huge problem during the extreme spring temperatures. In principle, local governments instruct farmers to provide potable water for their workers every two hours. In practice, however, child workers often resort to drinking water from irrigation or drainage canals. Even when farmers transport water directly to the fields, rather than drawing it from piped sources they simply truck in irrigation canal water instead. These are usually open canals that become vehicles for the distribution of human and animal solid waste and waste-borne pathogens. Children recruited to pick cotton near where they live are able to return home to their families in the evenings. But older children and those conscripted to work in the more remote cotton farms are forced to sleep in makeshift dormitories on farms or, ironically, in classrooms. Some children live in barracks with no electricity or windows. After weeks of arduous work and poor accommodation children are left exhausted and in poor health.

## WAGES

The monthly earning of child laborers is dependent upon various factors such as the nature of the job (skilled, unskilled, or semi-skilled work), duration of working hours, and nature of employment (daily wage, contract, or regular). Table 3.3 indicates the income of child laborers in the cotton cultivation of Uzbekistan.

The distribution of the monthly income of the child laborers brings out the fact that a large proportion of them, 103 (57.2 percent) in our study, earn between 1,200–1,600 Uzbek soms[6] (*сум*) per day and 45 (25 percent) earn between 600–1,000 soms per day on a given quota of cotton picking.[7] There were only 18 (10 percent) children who earned more than 1,700 soms per day. All the children were receiving their income in cash, some on a daily basis, some weekly, depending upon the type of work they were doing. Quality and quantity of work were the criteria that determined the wages, according to the majority of the child laborers, their parents, as well as the employers interviewed.

While it is certain that the Uzbek regime exploits children in forcing them to take part in the annual cotton harvest, it is less clear how much the children are paid. Some claim to receive US$5 for five days' work. Others report receiving just 15 US cents for the same period of labor. In 2001, the Uzbek NGO, *Tahlil*, estimated that payment for one kilogram of cotton ranged from 1.5 US cents at the beginning of the season to 1.0 US cent at the end. In 2004, children in the Fergana region reported that an average day's harvest of ten kilograms of cotton would earn them 38 US cents. For the Uzbek regime, forced child labor is undeniably cheap and immensely profitable. A child may be paid, at best, 3–4 US cents per kilogram for a commodity that is valued at US $1.15 on the global market.

Children usually receive some payment for their labor in the fields, although there are no standard rates that employers or state officials are required to meet. However, it is clear that farm administrators sometimes invent reasons to dock or to refuse to pay out promised funds, in full or in part. Sometimes farmers on the brink of bankruptcy pay the children in part, and the kids have no chance of seeing the rest of the money. School administrators often arrange for farmers to make payments directly to schools in cash with further payments to students in the future.

**Table 3.3**

**Table 3.3. Daily Earnings of Child Laborers**

| S. No. | Daily Earning | Number | Percentage |
|---|---|---|---|
| 1. | 1200 soms to 1600 soms | 103 | **57.2** |
| 2. | 600 soms to 1000 soms | 45 | **25.00** |
| 3. | **1700 soms and above** | **18** | **10.00** |

However, sometimes the farmer pays the children directly in cash so that they may be assured their due payment is complete. At times arrangements specifically exclude payment, when the children are brought out for a khashar for one day, or a few days, and are not paid anything because of a deal struck with the teacher or administrator.

Whatever little the child laborers earn they have to use it to pay for their food and lodging in run-down barracks. The responses of the child laborers in the field study also revealed that they were spending their earnings on personal and domestic needs if any money was left over after paying for food and lodging for themselves. No need is fulfilled through this little earning, according to a good percentage of the sampled child workers. A significant number of children said their contribution improved the overall economic conditions of their area and of the country as they considered it their duty to their country to pick cotton. But there were also a considerable number of children who were not in a position to improve their family life by their contribution, as they were earning little. According to them, they could hardly meet their personal requirements by their earnings.

## HEALTH CONDITIONS

Most of the work the children do is monotonous, repetitive, and dull and is often not suited to their physical and mental capabilities. Some children are ill-treated, humiliated, and even beaten, while others are looked after with parental care, which acts as an incentive and motivates these young children to undertake arduous work beyond their capacity for long durations. This adversely affects their health and well-being. It is against this backdrop that we asked children in the cotton industry of Uzbekistan about the working conditions in the handicraft centers.

The nature and amount of work have their effects on the physical as well as mental growth of the workers, especially the young ones. Cotton picking in particular requires intensive repetitive use of the fingers for picking the cotton from the plants and standing up for hours together. Thus, irritation of the eyes, fingers, and joints; back pain; and stomach and chest pains caused by inhaling cotton and dust can be said to be common. The child laborers in our study said there was no protection against work hazards. Table 3.4 and chart 3.1 shows the working conditions and health problems of working children.

Problems due to long working hours, hard and hazardous jobs, poor diet, and so on puts stress on the children. They complained of backache, headache, eye irritation, and pain in joints. Space for work is open, but it is too hot to be out in the open sun for such a heavy workload. Chronic and lifelong diseases emerge in this situation, and the government officials as well as the parents do not take it seriously. Our observation

Table 3.4

**Table 3.4. Health Problems of the Children in the Cotton Industry**

| S. No. | Health Problems | Number | Percentage |
|---|---|---|---|
| 1. | Back pain | 94 | **52.2** |
| 2. | Joint pain | 126 | **70.00** |
| 3. | Respiratory problem | 86 | **47.7** |
| 4. | Leg pain | 41 | **22.7** |
| 5. | **Stress/strain** | **72** | **40.00** |

reveals that the employer is not concerned about the health problems and safety measures.

The Paris-based group Human Rights in Central Asia reported that under the Soviet Union, forced labor was accompanied by some care for the health of children, the quality of their nutrition, and development of the rural social infrastructure. Under the new regime, children work at least eight hours daily on the cotton fields, sometimes without rest for days. They inhale dust, laden with residues of chemicals, pesticides, and defoliants. Moreover, the students are beaten up by school staff for refusing to work for the cotton harvest. Human Rights Organizations confirmed eight deaths of children working in the Samarkand region over a two-year period. Many more suffer with chronic diseases including intestinal infections, respiratory infections, meningitis, and hepatitis. Heatstroke is also common. The Convention on the Rights of the Child provides that children have a right "to be protected from performing any work that is likely to be hazardous, or to interfere with the child's education, or to be harmful to the child's health or physical, mental, spiritual, moral or social development."

In the opinion of local doctors, independently dangerous conditions collectively contribute to the children's weaker immunities and leave them more vulnerable to infectious diseases to which they are exposed to due to lack of elementary sanitation. Local medical opinion ties rates of viral hepatitis infections to pesticide and nitrogen fertilizer use. Research divulges that frequently this disease is contracted during spring fieldwork. Fatal outcomes are not uncommon. Amoebic dysentery and gastroenteritis are also prevalent. The cost of medical treatment is almost prohibitive, forcing families into debt. Farmers are not obligated to provide such treatment for those injured or sickened while working in their fields, nor are they required to provide any compensation to families for injuries or fatalities.

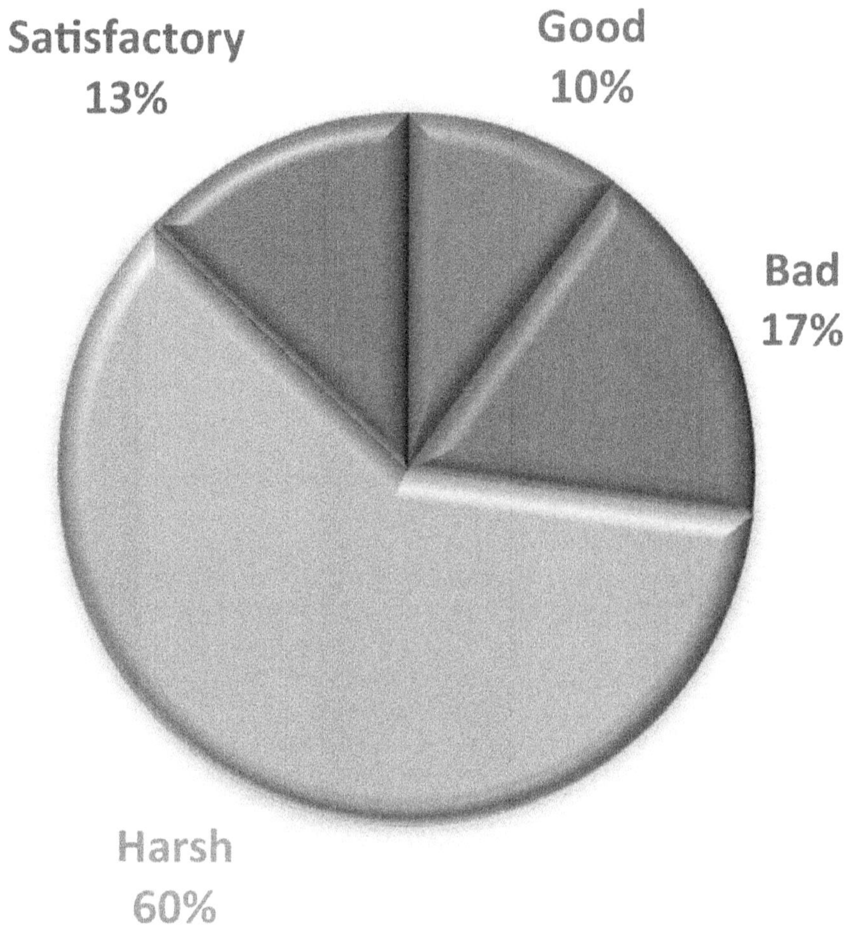


**Figure 3.1.**

CAUSES FOR EMPLOYING CHILDREN

The compulsory mobilization of schoolchildren for the cotton harvest represents a distinctive pattern that breaks with worldwide trends. Whereas families and employers tend to be both the major initiators and beneficiaries of children's work elsewhere, Uzbekistan represents a rare instance of state-sanctioned mass recruitment of child laborers. The various studies conducted by different agencies and organizations in Uzbekistan show that the recruitment of child laborers relies on a well-orchestrated nationwide campaign that involves foresight, planning, and coordination among public agencies on many different levels. Instructions are transmitted from local administrations, oblast and district *hokimiyats,* to local schools and farmers who are allocated a certain contingent of children. The latter have to accept the stated numbers of schoolchildren, to provide transportation, to create adequate conditions for work and to make timely payments. Local *hokimiyats* call daily meetings (the so-called *shtab*) where all administrators and farmers concerned report on the progress of the harvest. Central and local administrations engage in forward planning and take necessary measures for the allocation of resources; transport, fuel, medical assistance and cash, to ensure the efficient employment of labor during the harvest period. Clearly, these preparations involve the allocation of scarce resources and the license to suspend schooling for a lengthy period of time. Given the extensive presidential powers over the appointment and vetting of local administrators (spelt out fully in Article 93 of the Constitution) it is not conceivable that local *hokims* could take such initiatives without the tacit support or endorsement of the central government. Nor is there any evidence that the central government is using its extensive powers to take local administrations to task over their use of child labor. This suggests that the practice of compulsory child labor in the cotton sector of Uzbekistan is the result of public policy despite the fact that the government is a signatory to ILO Conventions that prohibit this practice.

## NOTES

1. Several studies document the mechanisms new farmers use to achieve better terms for their leaseholds by negotiating more profitable crop mixes and avoiding the unprofitable cotton crop. See Tommaso Trevisani, "The Emerging Actor of Decollectivization in Uzbekistan: Private Farming between Newly Defined Political Constraints and Opportunities," in D. Kandiyoti ed., *The Cotton Sector in Central Asia: Economic Policy and Development Challenges* (London: SOAS, University of London, 2007)..

2. There are some indications that the pressures created by the diversion of land resources are being countered by administrative measures. The "grave shortcomings" leading to the sackings of the acting governor of the Tashkent Region, three district heads and one mayor include "the embezzlement and illegal appropriation of state agricultural lands for private use." See "Uzbekistan: A Purge Is Underway," *Eurasianet*, December 17, 2008, http://www.eurasianet.org.

3. Government of Uzbekistan, *Welfare Improvement Strategy Paper*, 2007, p. 21.

4. According to Russian Central Bank figures, migrants from Uzbekistan transferred US$1 billion in 2006. However, it must be borne in mind that many also use informal channels for money transfers. *Ferghana News*, http://www.ferghana.ru/news.

5. According to some reports, although the current economic downturn is translating into a lower volume of remittances, there is no decline in the number of Kyrgyz, Uzbek and Tajik citizens seeking jobs abroad. See Erica Marat "Shrinking Remittances Increase Labor Migration from Central Asia" *Central Asia-Caucasus Analyst* 11, no. 3 (February 11, 2009):7–9.

6. One US dollar is equivalent to 1,665 Uzbek soms (*сум*) officially, but the exchange rate of dollar and som in the market is higher, around 2,260–2,460 soms per dollar in 2010–2011, when I was doing the field study in Uzbekistan.

7. Child workers in cotton cultivation are required to pick between 100 to 176 pounds (80 kgs) a day. For each kilogram of cotton the workers pick they are paid 15–20 Uzbek soms, or 1–2 cents.

# FOUR

# Social and Legal Aspects

While education itself is clearly significant for the development of an individual and for the well-being of the society, it may be less obvious why education is important in the context of child labor. Widespread child labor is closely associated with poverty. The family may depend on the contribution a working child makes to the household income, and place more importance on that income than on education. A poor family may be unable to afford school fees, uniforms, or other costs. Moreover, when a family has to make a choice between sending either a boy or girl to school, it is often the girl who loses out. While child labor is often a much needed contribution to the income of the family, education may be so costly that the decision whether to send a child to school or to work will be difficult for many poor parents.

## CHILD LABOR AND SCHOOL ACHIEVEMENT

A child who is in school full-time is more likely not to engage in child labor. Hence, expanding access to free and compulsory education is crucial to reducing child labor, as is the provision of quality education. Access to education is a necessary but not sufficient element, as the challenge is to retain children in school. Only quality education can ensure that they stay in school. Quality means that teachers are recruited in adequate numbers to avoid high student-teacher ratios in classrooms. Teachers and educators need to receive the training required to make them effective. Relevant curricula are also essential for an education of quality. Finally, no good education can be provided if classroom conditions are deplorable and students lack the necessary books, equipment and other educational materials. Children who receive quality education are more empowered to escape from poverty and, as adults, are more likely to send their children to school.

Myron Wiener and others hold the view that the universal extension of state-funded education in Europe, North America, and Japan has been the most powerful instrument for the abolition of child labor. "No country has successfully ended child labor without first making education compulsory. As long as children need not attend school, they will enter the labor force."[1] Policy makers in most countries believe that mandatory education is a prerequisite for the eventual abolition of all forms of child labor. Wiener refuted the popularly accepted explanations given by scholars on child labor, viz., low economic development of the nation, poverty, high birth rate, and size of family.[2]

The expected return to education is an important factor in the parents' considerations. However, the true benefits of schooling, such as increased future wages may often not be known to families, and even if they are, weighing up value against private costs is not an easy task. Even if parents are aware that the returns on education could be significant, the cost of schooling can be so high that the children are removed from school and pushed into works.[3] The benefits of going to school are mostly long term, and will mainly affect the child, rather than the parent. On the other hand, the costs have to be borne by the parents, and these costs are due in the short term. A comparative study of child labor and schooling in Africa found that one way to reduce child labor and increase incentives to keep the children in the educational system is to improve access to credit, which the family can pay back later.[4]

Education of good quality is a key element in the prevention of child labor. Children with no access to education have little alternative than to enter the labor market. Child labor is also one of the main obstacles to education for all, since children who are working full-time cannot go to school. The academic achievement of children who combine work and school often suffers, and there is a strong tendency for these children to drop out of school and enter into full-time employment.

When education is compulsory and attractive, it can help to reduce child labor. There is need to develop an integrated policy and program of action to provide quality free and universal education that is relevant and accessible to children in poor families such as those to which the majority of child workers belong. Child labor concerns should be explicitly addressed and integrated into such a policy and program of action; a holistic approach is necessary. Quality education should be provided for children at least up to fifteen years of age to stem the flow of children into the labor market and to provide working children and former working children with educational opportunities. A recent study estimates that worldwide an additional US$16 billion per year would be required to achieve universal primary education of decent quality by 2015.[5]

The table 4.1 shows that the world can surely afford to invest more money in the well-being of its children. However, while universal access to education is affordable in global terms, there are many countries, particularly in Africa, for which the costs of financing education programs are too high. These countries must rely on increased donor assistance from the developed countries to meet the present educational challenges.

NGOs can play an important role in this regard. It is imperative to link education with the individual's life, need, and ambitions so as to make him a powerful instrument of social, economic, and cultural transformation, and to fulfill national objectives through him. This is possible only when education is linked with productivity. Simultaneously it can develop a truly indigenous life style along with modernization. Above all, education can develop a sense of cooperation and can formulate character through stress on social, moral, and spiritual values.[6]

## EDUCATIONAL LOSSES

Education is of paramount importance for the proper growth and development of the individual. It plays an important role in one's life in the sense that it helps in shaping the right kind of life style in human beings. Education, formal or informal, aims at developing the positive quality of the child and helps the individual realize his potentialities. All the children working in the cotton sector that were interviewed were studying in the schools and Lyceums (colleges) in different grades. Accordingly, the child workers were asked to state their educational qualification/background.

Almost one-third of the interviewed children, 58 (32.2 percent), had an educational qualification of fifth grade only; 72 (40.00 percent) had an

**Table 4.1**

**Table 4.1. Annual Expenditure on Basic Education and Other Items**

| No. | The World's Priorities (annual expenditure) | |
|---|---|---|
| 1 | Basic Education | 6 bn |
| 2 | Ice Cream in Europe | 11 bn |
| 3 | Perfumes in Europe and USA | 12 bn |
| 4 | Pet foods in Europe and USA | 17 bn |
| 5 | Business entertainment in Japan | 35 bn |
| 6 | Cigarettes in Europe | 50 bn |
| 7 | Alcoholic drinks in Europe | 105 bn |
| 8 | Narcotic drugs in the world | 400 bn |
| 9 | Military spending in the world | 780 bn |

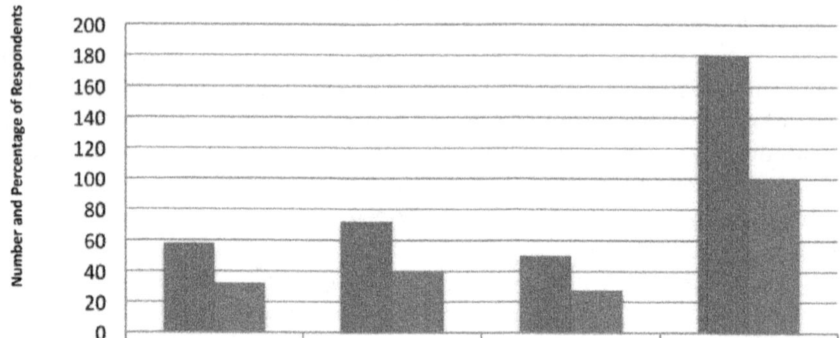


| | upto 5th grade | upto 8th grade | upto 10th grade | total |
|---|---|---|---|---|
| Number | 58 | 72 | 50 | 180 |
| percentage | 32.2 | 40 | 27.7 | 100 |

**Figure 4.1.**

educational qualification of eighth grade; and the remaining interviewed children, 50 (27.7 percent), working in the cotton cultivation of Uzbekistan were students of the tenth grade. Children mobilized for the cotton harvest experience significant educational losses. Two months of school closure during the harvest period are de facto deducted from their schooling. This represents a net loss of about 25 percent of their exposure to education. The cumulative effects of these losses over the years, between grades five to grade nine, amount to one entire year of net school time. Other field reports suggest that schoolchildren in rural areas may also be subject to recruitment for spring farm labor (hoeing, weeding, and transplanting) occasioning further suspension of classes for weeks at a time. Thus, the total educational losses of rural schoolchildren are even higher than those identified by various studies. Many schoolchildren are acutely aware of the significance of this loss for their future prospects. They are at a clear disadvantage in comparison to their urban counterparts and are resentful of these obstacles to their educational and social mobility. Some interviews with children, their parents, and teachers express concern and discontent with the current state of affairs.

## LEGAL ASPECT

Human rights were formally acknowledged and declared in realization of the need to safeguard and promote the inherent dignity of all members of human society. Prior to the Convention on the Rights of the Child, human rights standards applicable to all members of the human family had been expressed in legal instruments such as covenants, conventions, and declarations. By, 1989, the standards concerning children were brought together in a single legal instrument agreed to by the international community. It unambiguously spelled out the rights to which every child is entitled, regardless of place of birth, descent, sex, religion, or social origin. Uzbekistan has always been a land of many social problems and one among them is child labor in the cotton industry. The Constitution of Uzbekistan guarantees fundamental rights and full freedom for the children of the country to experience childhood. In spite of that, millions of children are being put to arduous work for little gain.

## CONSTITUTION OF UZBEKISTAN

In most societies, the constitution is the basic legal document that determines the relationship between the people and the state, the form of the government, and the conditions under which to adopt or suspend the basic rights and freedoms. After acquiring its independence, the main aim of legislative activity in Uzbekistan was the acceptance of a new constitution.[7] The constitution itself contains several democratic principles and it is more democratic than the constitutions of some of the so-called democratic states. It remains, however, only a document at the present time (most presidential decrees become law). The electoral system in Uzbekistan is based on proportional representation. The old constitution, adopted in 1978, enshrined the Communist Party as the leading and directive power in society. It consolidated the priority of the state's nationwide ownership as the basis of the economy. After independence, these ideas were fetters on sociopolitical and economic transformation. On December 8, 1992, the Supreme Council of Uzbekistan adopted a new constitution. The authorities claimed that it had been adopted after a nationwide discussion that established the basic principles of the social system, and that it consolidated the basic freedoms, rights, and duties of citizens. An examination of the major newspapers published in Uzbekistan shows that there were virtually no dissenting voices over any of the constitution. This demonstrates that none of the opposing views were given expression in the mass media before the referendum.

The new constitution of 1992 defines Uzbekistan as a sovereign, democratic, secular, and presidential republic. All citizens of the Republic of Uzbekistan, regardless of their ethnic origin, constitute the people of Uzbekistan. The constitution declares the people to be the only source of the government, and the state shall express the will of the people and serve their interests. Public life is to develop on the basis of the diversity of political institutions, ideologies, and opinions. State bodies and officials shall be accountable to society and the people. Priorities of personal, economic, political, and social freedoms and rights of citizens are clearly stated in the constitution. It also devotes itself to participation by the citizens in the political process and is committed to a principle of individualism that is unusual in Central Asia. The course of development on the domestic policy scene is, however, going in the opposite direction. Political pluralism is associated with destabilization, "street democracy," or anarchy, and the opposition is oppressed under the guise of "maintaining stability" in a region threatened by instability.

The system of state authority in the Republic of Uzbekistan is based on the principle of the division of powers between the legislative, executive, and judicial authorities. The legislative authority belongs to the parliament (Oliy Majlis), the executive authority belongs to the president and the Cabinet of Ministers, and the judicial authority belongs to the courts, led by the Supreme Court. The constitution also determines the territorial administrative system of the republic. According to the constitution, the Republic of Uzbekistan consists of regions, districts, cities, towns, *kishlaks* and *auls* (villages) and the autonomous Republic of Karakalpakistan. It also states that the sovereign Republic of Karakalpakistan is part of the Republic of Uzbekistan and its sovereignty is protected by the Republic of Uzbekistan. The Republic of Karakalpakistan is to have its own constitution, which must be in accordance with the constitution of Uzbekistan. The laws of the Republic of Uzbekistan are binding in the territory of Karakalpakistan. Any alteration of the boundaries of Karakalpakistan must be sanctioned by the Oliy Majlis of Uzbekistan, but this may not be done without the content of Karakalpakistan.

The constitution determines and consolidates the fundamentally new system of state body authorities in regions, districts and cities where *hokims* are the leaders of both local representative bodies and executive authorities. They carry out their tasks on the basis of one factor—that they are not elected but are appointed (or dismissed) by the president. Thus a strong executive authority has been formed and led by the president. The political elite of Uzbekistan claims that the constitution of 1992 is the work of a profound study of the experience of constitutional and democratic processes in countries with developed democracy, but at the same time a reflection of specific Uzbek conditions and traditions. Taking into consideration all the advantages of the new constitution, especially in comparison to the old one, we should remember that it determines only the general principles of the social system. A constitution, even the most democratic one, cannot provide democracy, freedom and a law-based society solely by itself. It serves only as a compass that point out the goals of social development. But it remains as a document if there is no political or economic freedom for citizens. If the conditions for implementing these freedoms are absent and if the legal and political culture is not sufficiently developed, then the constitution is simply a part of the process of democratic development, even if it is a very significant part. One should neither denigrate nor exaggerate its role. Social development itself will determine only the course of democratic development, and the constitution is just a reference point on this way.

## CONVENTION ON THE RIGHTS OF THE CHILD

The General Assembly of the United Nations adopted the Convention on the Rights of the Child (CRC) on November 20, 1989. It stands for the protection of children everywhere against exploitation, neglect, and abuse.[8] The Convention draws attention to four sets of civil, political, social, economic, and cultural rights of every child. These include (1) The Right to Survival; (2) The Right to Protection; (3) The Right to Development; and (4) The Right to Participation. Article 23 of the Convention says, "All States should work to end such child labor practices and see how the conditions and circumstances of children in legitimate employment can be protected to provide adequate opportunity for their healthy upbringing and development."[9]

## WORLD CONFERENCE ON HUMAN RIGHTS

The World Conference on Human Rights organized by the UNO, which was held in Vienna, Austria, from June 14–25, 1993 deeply, concerned the rights of Children in global society. The conference reiterated the principle of "First Call for Children"[10] and, in this respect it underlined the importance of major national and international efforts, especially those of the United Nations Children's Emergency Fund (UNICEF), for promoting respect for the rights of the child to survival, protection, development, and participation. It called on states to integrate the Convention on the Rights of the child into their National Action Plans.[11]

The World Conference on Human Rights urged all states, with the support of international cooperation, to address the acute problem of children under difficult circumstances. Exploitation and abuse of children should be actively combated, by addressing their root causes. Effective measures are required against harmful child labor. The conference supported all measures by the United Nations and its specialized agencies to ensure the effective protection and promotion of human rights of the female child. It urged states to respect existing laws and regulations and remove customs and practices that discriminate against and cause harm to the female child. It recognized the important role played by NGOs in the effective implementation of all human rights instruments and, in particular, the Convention on the Rights of the Child.[12]

As early as 1921, when the International Labor Organization (ILO) passed the first Minimum Age Convention, the world has attempted to protect children's right to an education and to prevent any child labor that would prejudice their school attendance.[13] The ILOs Minimum Age

Convention 138 of 1973[14] set the standard for the minimum age for admission to employment at fifteen years, or in special cases where economic and educational facilities are insufficiently developed, fourteen years; light work not harmful to the child or prejudicial to his or her attendance at school is permissible after age twelve. Since 1990, with the entry into force of the Convention on the Rights of the Child,[15] the child's right to be protected from "any work that is likely to be hazardous or to interfere with the child's education" (Article 32) and his or her right, on an equal, nondiscriminatory basis to "primary education compulsory and available free to all" (Article 28) have gained the status of internationally recognized norms, while imposing an obligation on the 192 states parties to the Convention to realize these rights for the children under their jurisdiction.[16] In 2000, children were provided further protection through the entry into force of ILO Convention 182,[17] which was ratified by 150 countries as of May 2004.[18] Convention 182 prohibits the worst forms of child labor, defined as all forms of slavery and similar practices; child prostitution and pornography; illicit activities (in particular the production and trafficking of drugs); and work that is likely to harm the health, safety, or morals of children.[19]

The Geneva declaration of the Rights of the Child 1924 was the first convention of League of Nations to discuss the rights of child.[20] The Universal Declaration of Human Rights as adopted by the United Nations in 1948 incorporated the basic rights of children for growth and education. An independent "Declaration on the Rights of the Child" was made by United Nations in 1959.[21] This emphasized special protection and care for children to develop in a normal and healthy manner in an atmosphere of freedom and dignity. The United Nations Declarations emphasized child welfare, having applications to every social condition by conferring specific rights to children through social legislation prohibiting any discrimination.[22] The year 1979[23] was designated as International Year of the Child (IYC) by the United Nations, which stressed the cooperation of nations in common tasks of meeting the basic needs of the children—nutrition, health, education, maternal protection, family care, equal society status, and protection from racial and other forms of discrimination.[24] This was a challenge to the conscience of mankind and to the community of nations.

On November 20, 1989, the General Assembly of the United Nations adopted a Convention of Rights of the Child that was ratified by 107 member countries within a year. The convention contains fifty-four articles, each dealing with different types of rights. These can be broken down into four broad categories such as survival rights, developing rights, protection rights, and participation rights. Several rights cover a child's right to the basic needs for existence, which include adequate living standard, shelter, nutrition, and access to medical services. Developing rights include those that help children to reach their fullest potential, such as rights to education; play and leisure; culture activities; access to information; and freedom of thought, conscience, and religion. Protection rights require that children be safeguarded against all forms of abuse, neglect, and exploitation, including child labor, drug abuse, sexual exploitation, abuses in the criminal justice system, and so on. Protection rights allow children to take an active role in their communities and nations. These encompass the freedom of expression, to join associations, and to assemble peacefully, which is necessary for preparation of a complete and responsible adulthood.[25]

In 1919, the International Labor Organization, in its first session adopted a convention on the minimum age of children for admission into industrial employment. It said that no child under fourteen should be employed in any public or private undertaking other than ones in which only family members are employed.[26] Industrial undertaking here includes mines, quarries, manufacturing units, construction, maintenance, and repairs, as well as transportation of passengers and goods by road, rail, or waterways. The minimum age was revised from fourteen to fifteen in the Convention No. 59 of 1973. The Convention which grew out of the personal initiative of the Director-General Wilfred Jenks,[27] marks important progress in raising the standard for general employment in developed countries to fifteen years or the age at which compulsory education ends, whichever is the higher (accepted in 1973).[28] Subsequently the ILO prohibited employment of children in agriculture, shops, hotels, restaurants, offices, places of public employment, and others. Convention No. 138 of 1973 of the ILO emphasized that each member for which the convention was in force would undertake to pursue a national strategy designed to ensure the effective abolition of child labor and to raise regressively the minimum age for admission of children to employment. The same convention also recommended special attention for working conditions of young persons under eighteen years, in terms of fair remuneration, limited working hours, prohibited night work, weekly and annual holidays with leave, coverage by all social schemes, and maintenance of satisfactory standards of health and safety.

ILO has adopted five conventions for making medical examination of young persons a condition precedent for employment. This is to ensure fitness of the young persons for the employment where he/she is to be employed.[29] Also, three conventions were made in support of prohibition of night work of young persons both in industrial and nonindustrial establishments.

Child agricultural labor violates local and international laws and norms. The Government of Uzbekistan is a signatory of numerous international human rights and labor treaties. It has the necessary legal framework to eradicate child labor. Nonetheless, in terms of international law, child labor in the cotton sector of Uzbekistan contravenes several articles of the 1989 *Convention on the Rights of the Child.* Article 28, paragraph (e) affirms that State Parties recognize the right of the child to education, and shall, in particular "take measures to encourage regular attendance at schools and the reduction of drop-out rates." It is evident that schooling is being disrupted for two months a year for the cotton harvest and that additional disruption sometimes occur for spring farm work (such as hoeing, weeding and transplanting). Article 31declares "the right of the child to rest and leisure." Working children in Uzbekistan are deprived of this right during the harvest period. They have a full working day without weekend breaks. Transportation to and from the cotton fields adds to the length of the working day and some children may not return home until late in the evening without the benefit of rest and recreation. Finally, Article 32of the Convention affirms the "right of the child to be protected from economic exploitation and from performing any work that is likely to be hazardous or to interfere with the child's education." The realities of child labor in the cotton sector of Uzbekistan, suggest that this norm of international law is also being contravened. The use of child labor to harvest cotton also violates the ILO conventions to which the government is a signatory.

Uzbekistan has signed and ratified ILO conventions: No 29 (the 1930 Forced Labor Convention) and No. 105 (The 1957 Abolition of Forced Labor Convention). These documents were ratified in 1992 and 1997 respectively. However, neither aforementioned national laws nor Uzbekistan's ILO commitments have curbed widespread use of forced child labor. Until 2008, two other important ILO conventions, No. 138 (Minimum Age Convention, 1973), and No.182 (Worst Forms of Child Labor Convention, 1999) were unrecognized by the Uzbek government. Finally, in March 2008, the Parliament (Oliy Majlis) ratified these two conventions. In September 2008 the Cabinet of Ministers adopted a resolution and the National Action Plan aimed for the implementation of C182 and C138. But these documents didn't stipulate cooperation with ILO or any mechanism of independent monitoring of how the conventions were being implemented. A few weeks after the resolution, schoolchildren were, as usual, taken for cotton harvest. That fact demonstrated that the rule of law is nonexistent in Uzbekistan, and none of the ILO conventions ratified by Uzbekistan are considered seriously by its government. None of them have affected the real situation. Nor has the ratification obviated the need for further pressure on the Uzbek government to end the use of forced child labor.

## NATIONAL LEGISLATION

In terms of domestic law, child labor in Uzbekistan contravenes several items of national legislation. Article 37of the Constitution prohibits the use of any form of forced labor. The 1991 *Law on the Foundations of State*

*Youth Policy* stipulates that "it is not permitted to attract school and university students to work during the learning process, except in cases when such work corresponds to their chosen specialty and is a form of apprenticeship, or cases of voluntary collective or individual work in time free from schooling. Such labor is accepted on the condition of properly concluded contracts in accordance to labor and civic laws." The Law on *Guarantees of the Right of Child to Labor* adopted in 2007 states that a person younger than eighteen years of age is considered to be a child. The right to employment may be exercised from sixteen years onward, and in some cases (with the consent of parents and during periods free from study) from fourteen years on. At the same time, the state guarantees the labor rights for persons younger than eighteen years by providing the necessary conditions for combining work with compulsory education (Article 20). Article 20makes clear stipulations concerning the conditions under which children may be permitted to work including (1) Every child has the right to work, free choice of the type of activity and profession, fair labor conditions in accordance to his age, state of health, and professional training in accordance with the procedure prescribed by the law. (2) Application for a job is permitted from the age of sixteen. (3) Persons who reach the age of fifteen can be accepted for a job by written consent of either parents or guardians. (4) To prepare the children for work it is permitted to recruit the pupils of secondary schools, academic lyceums, and professional colleges to carry out an easy job, that does not harm their health and growth, does not infringe upon the educational process, is free from school hours—upon reaching the age of fourteen by written consent of either of parents or guardians. Article 7of the Labor Code of the Republic of Uzbekistan on "Prohibition of forced labor" clearly states that forced labor, namely a compulsion to work under the threat of any punishment (including as a means to ensure labor discipline) is prohibited except when it is executed on the basis of legal acts on military or alternative services, in a state of emergency, as a result of a court verdict coming into force or other cases envisaged by the legislation. Article 241of the Labor Code, also prohibits the use of children for any work that may damage their health, security, and morality. The government of Uzbekistan points out that these provisions are in full compliance with the international legal acts on protection of children's rights and, in particular, the 1989 *Convention of the Right of the Child*. However, there is a widespread international concern in relation to the actual implementation of these laws.

The research delineates that the application part of Uzbek legislations and provisions related to the safeguard and rights of children from forced labor and other forms of exploitation are unimplemented, as the state is the major agency breaking the laws they formed. Children aged between six to eighteen years are put to arduous labor for narrow and short gains of the autocratic state. In this way the state is putting its own future in stake as the children are the future of any state. Far from implementing the legislations and legal provisions related to forced labor; the state of Uzbekistan is in itself its violator. Islam Karimov's government is also not allowing human rights and other nongovernmental organizations to work in Uzbekistan for the welfare of the children working under conditions damaging to their health and education and are not allowed to highlight the pathetic condition of children in different sectors of the economy, especially the cotton industry.

The national as well as international provisions, legislations, conventions, articles, and constitutional safeguards are hardly visible in the field of working children. The poor implementation of the constitutional measures for the safeguarding of the children's rights including education and health rights especially is an outcome of the state's dependence on the monoculture of cotton.

## LEGAL PROVISIONS AGAINST CHILD LABOR IN UZBEKISTAN

Uzbekistan has adopted a number of laws, on Youth Policy, on Labor, and on Children's Rights. These statutes set limits and restrictions on the use of child labor. Legally, children are only allowed to work after age fifteen. Before this age they can be employed only in the context of school-related activity, and for no more than fifteen days.[30] In all cases, children are not allowed to work more than four hours per day.

Article 37 of the Constitution of Republic of Uzbekistan guarantees all citizens the right to work, in appropriate conditions, and to choose their employment freely. The social and labor rights of citizens are contained in two basic documents—the Labor Code (1995) and the Law "On Employment of the Population" (1992, revised in 1998). According to the Law "On Employment," all citizens have "equal opportunities in the realization of their right to work and to a free choice of employment, irrespective of sex, age, race, nationality, language, social origin, property and official position, religion, membership in public associations, or other circumstances that have no negative impact on the ability of persons to work and the results of their work" (Article 5).[31]

According to Article 77 of the Labor Code, children have the right to be employed at the age of sixteen, when they finish their primary education (nine grades), which is compulsory under the Law "On Education." Also, all male citizens between the ages of sixteen and sixty-four (sixteen to sixty for women) are considered able-bodied and thus, part of the labor force. According to international law, all persons under eighteen years are minors, which is compatible with the Civil Code of Uzbekistan which stipulates that children attain maturity at eighteen, when they are granted the right to vote and run for public office.

## UZBEKISTAN LEGISLATION AND INTERNATIONAL HUMAN RIGHTS COMMITMENTS

Economic exploitation is prohibited in both the Constitution and Labor Code of Uzbekistan. Namely, Article 37 of the Constitution of Uzbekistan prohibits forced labor, except as punishment under the sentence of a court, and the Labor Code of Uzbekistan sets sixteen as the minimum age for admission to employment, although children aged fourteen are permitted to work after the hours of study in light work, as long as it poses no hazards to their health or moral development, with the permission of a parent or guardian.[32] Article 241 of Uzbekistan's Labor Code prohibits the use of child labor, "which can damage [their] health, safety and morality." Article 8 of the Labor Code states that "the Republic of Uzbekistan directly prohibits child labor during education periods, if such activity is not related to the major subjects or part of an internship, or if outside the education period based on an individual or collective volunteer action.

A joint legal resolution issued by the Ministry of Labor and the Ministry of Health, and registered by the Ministry of Justice in September 2001 (No. 1040), lists cotton picking and other forms of child labor on a national list of unfavorable working environments prohibited to workers less than eighteen years of age.[33] The rights of a child to education and to health are also stipulated in Uzbek laws "on Education" and "on Protection of Public Health." However these regulations are not enforced. In practice, the economic exploitation of school-age children is widespread and organized by the government itself.

The US Department of State's Human Rights Report for 2006 stated that the Government of Uzbekistan did not effectively implement laws and policies to protect children from exploitation in the workplace. According to the report: A 2001 government decree prohibits those under age eighteen from engaging in manual cotton harvesting and other jobs with unhealthy working conditions; however, in rural areas children often help to harvest cotton and other crops. The large-scale compulsory mobilization of youth and students to help in the fall cotton harvest continued in most rural areas. Such labor was poorly paid. There were occasional reports from human rights activists that local officials in some areas pressured teachers into releasing students from class to help in the harvest and in many areas, schools closed for the harvest.[34] The State Department's Trafficking in Persons report released in June 2007 also confirmed that men and women are trafficked for the purpose of forced labor in agriculture and designated Uzbekistan as a Tier 3 country.[35]

The main law enforcement agencies in Uzbekistan are the Prosecutor General and the Ministry of Interior. The Ministry of Labor and Social Protection does not have legal jurisdiction over child labor enforcement.[36] The laws provide for criminal and administrative sanctions to punish violators of child labor laws. However sanctions were not adequate to deter violations related to the cotton harvest, and in any case were not enforced. There were no reports of prosecutions or administrative sanctions resulting from any inspections.[37]

Uzbekistan is also a party to numerous international human rights and labor rights treaties. Since its independence in 1991, Uzbekistan has ratified all six major UN international human rights treaties, including the Convention on the Rights of the Child. Article 32 of this Convention recognizes the right of a child "to be protected from economic exploitation and from performing any work that is likely to be hazardous or to interfere with the child's education or to be harmful to the child's health or physical, mental, spiritual, moral and social development."

But the ground realities are quite a contrast. It was evident by the observations made by the UN Committee on the Rights of Child in April 2006. In its operational past, the statement goes: "We are deeply concerned at the information about the involvement of the many school age children in the harvesting of cotton resulting in serious health problems such as intestinal and respiratory infections, meningitis and hepatitis."[38]

Uzbekistan has ratified several International Labor Organization (ILO) Conventions, including Convention 29 on Forced Labor, and Convention 105 on Elimination of Forced Labor. However, it has failed to ratify Convention 138 on the Minimum Age for Admission to Employment of 1973 or Convention 182 on the Worst Forms of Child Labor of 1999.

In 2005 the Government of Uzbekistan submitted its first report concerning ILO Conventions No. 29 and 105 to the ILO Committee of Experts. The Committee has issued a Direct Request to the Government of Uzbekistan for more information concerning the use of forced labor. Unfortunately none of these documents are publicly available.[39]

In 2004, representatives from the Government of Uzbekistan participated in an assessment mission to gather preliminary information about the child labor situation in Central Asia.[40] Subsequently the US Department of Labor provided funding to ILO-IPEC for a sub-regional project to enhance the capacity of national institutions to eliminate the worst forms of child labor and to share information and experiences across the subregion.[41] However there has been no positive change in the situation in cotton harvesting in Uzbekistan since the initiation of that project.

The Government of Uzbekistan has officially denied the existence of forced child labor in cotton harvesting. In private, however, some Uzbek officials do admit the use of forced child labor in the cotton harvest.[42] In a 2004 interview with journalists from the Institute of War and Peace Reporting, an official from the Jizzakh regional administration privately admitted that the use of child labor was widespread and that the cotton industry couldn't survive without it.[43] On October 16, 2006, according to an Uzbek senior official, Cotton Fair in Tashkent collected exports contracts for 1.7 million tons of cotton fiber. However according to local informants interviewed by ILRF staff, lack of agricultural infrastructure and a corrupt subsidiary system discourage farmers from cultivating cotton. The potential social ramifications and economic inefficiencies have led the World Bank and the UNDP to urge the Uzbek government to change its agricultural policy, particularly in cotton cultivation. Until now the Uzbek Government has made little or no effort to liberalize the sector.[44]

In light of the serious violations of internationally recognized workers' rights identified in this petition, and the failure of the Government of Uzbekistan to undertake any meaningful steps to address these issues, it is clear that the country's benefits under GSP should be revoked. Uzbekistan's practice of using state-orchestrated forced child labor in national cotton production is a clear and substantial breach of its commitments under ILO conventions prohibiting forced labor. These practices are also clearly in conflict with ILO Convention 182 prohibiting the worst forms of child labor. Uzbekistan has neither ratified ILO 182 nor otherwise undertaken meaningful actions to prohibit the practice of forced child labor in the substantial and economically significant cotton sector.

## SOCIETAL ATTITUDE TOWARD CHILD LABOR

A less tangible but no less corrosive effect of existing labor practices in the cotton sector of Uzbekistan is the erosion of trust in the government's ability to deliver equitable development. There is, undoubtedly, a Soviet legacy of mobilization for cotton harvests, which used to be accompanied with propaganda and "socialist competition" among harvest brigades in order to push up production norms. However, cotton cultivators were also the beneficiaries of health, education, and welfare entitlements that increased their social wage. The current low levels of "trickle down" of cotton revenues to direct producers is not lost on the population. Xenophobic exhortations to participate in the production of national wealth under current conditions breed cynicism and discontent, which is further exacerbated by the exploitation of children.

The societal attitude toward children put to work in cotton fields is that it is inhumane and unjustified on the part of the state and administration. The modern civilized world has developed many tools and social indicators for measuring civilizations as developed, underdeveloped, and developing in socioeconomic terms. Perhaps, there cannot be a social indicator of civilization that treats its children as commodities.

The majority of the children, social workers, human rights activists, parents, and teachers interviewed were fully aware of the violation of local and international laws regarding the participation of children in cotton cultivation in Uzbekistan. We were allowed interaction with a limited number of local low-level officials, and only few who were interviewed denied that children were forced to pick cotton in the fields, saying, "If there is any child in the field, you must understand that they are there voluntarily." However they admitted having set up special agricultural offices within the prosecutor's office to pursue cases against those who will not fulfill quotas or other obligations, including schoolchildren.

In spite of the awareness, some of those interviewed suggested a number of causes for the necessity of children's participation, the most important being farmers' lack of means to engage farm machinery or pay adult laborers. It is always too expensive to pay adults. They demand defined working hours, respect for their rights. If you don't satisfy their demands, they won't work. Therefore local governments and farmers find it convenient to send children out to the fields. They are easy to handle and earn low wages compared to adults. Absurdly low purchase prices for cotton and wheat often have farmers unable to cover basic costs.

One of the farm directors in Samarkand delineated that beyond the insurmountable financial cost, the sheer lack of able-bodied adults in the villages is often put as a reason for the forcible recruitment of children (though if prevailing wages on Uzbek farms were higher, as they are in Kazakhstan and Kyrgyzstan to which increasing number of rural Uzbeks go to work, migration from the countryside could decline or reverse itself).

## ROLE OF NONGOVERNMENTAL ORGANIZATIONS

National and International nongovernmental organizations, can and do play a major role in providing vital services for deprived groups in publicizing child abuse and exploitation and in promoting the interests of the working children. Their role is all the more critical in situations where freedom of association and expression is discouraged and limited and where the dimension and severity of child exploitation are either unrecognized or deliberately concealed. Depending on the national situation within which they are operating, nongovernmental organizations can carry out a number of activities. They can initiate and facilitate the organization of especially vulnerable groups of workers. There are several nongovernmental organizations that have successfully organized communities, especially deprived groups, and helped them to protect and promote their interests. This wealth of experience could also be used to facilitate the delivery of social and economic services and to improve the socioeconomic status of low-income groups, which in turn could lead to a reduction in the incident of child labor. They can also carry out extensive educational and information campaigns on the evils and consequences of child labor and can encourage positive action locally, nationally, and internationally. They have been playing advocacy role but can also directly identify work situations in which cruelty to children and abusive practices occur. They may also, either individually or in cooperation with other interested groups, establish a mechanism for monitoring and reporting on such inhuman practices.

The role of NGOs in the contemporary times in different sectors of the social and economic structure of the society is critical and noteworthy. The organizations especially working in the field of human rights in many developing countries have a very hard task on their hands as these countries are under very unstable political and economic structure. Uzbekistan with its autocratic political structure presents a stiff challenge to nongovernmental organizations in various sectors including the civil and human rights.

UNICEF, the largest world organization dealing with the rights of the child world over delineated through its officials that there has been very limited success in curbing child labor in Uzbekistan. A recent panel discussion at the Open Society Institute highlighted the ongoing problem of forced child labor in Uzbekistan and the efforts of nongovernmental groups to enlist governments and international institutions in the cause of persuading Uzbekistan to end the practice. [45]

Umida Niyazova, an Uzbek emigrant and leader of the Uzbek-German Forum for Human Rights, described worrisome new trends involving increased police presence and harassment of human rights monitors who tried to track child labor. Niyazova's group verified that despite the Uzbek leadership's claims to the contrary, the practice of sending children into the fields has continued. At first, mainly older children were dispatched early in September, but then the recruitment of children as young as ten increased over the period of the harvest due to a shorter season with impending cold weather, and the high price of cotton on the world market this year after floods in Pakistan and China.

The Europe and Central Asia Division of Human Rights Watch have been generating and raising awareness about human rights concerns in Uzbekistan related to the use of young children in forced cotton picking and also about the restrictions on the human rights activists and media with the European Union. [46] UNICEF has been under fire from NGOs due to its failure to take up explicitly the issue of forced child labor in Uzbekistan and its use of an outdated survey that minimized the extent of the problem.

## NOTES

1. Myron Weiner, *The Child and the State in India: Child Labour and Education Policy in Comparative Perspective* (Princeton, NJ: Princeton University Press, 1990).
2. Babita Agarwal, *Child Labour in India* (Jaipur India: ABD Publishers, 2007), 60.
3. Save the Children, http://www.savethechildren.org.uk.

4. James W. Martin, Klaus Gunther, and Sule Caglar, *Child Labour in Europe and Central Asia* Working Paper (Geneva: ILO, IPEC, 2003).

5. International Programme for the Elimination of Child Labour, *Combating Child Labour Through Education* (Geneva: ILO, 2003).

6. Agarwal, *Child Labour in India*, 69.

7. Uzbekistan Respublikasining Konstitutsiyasi [Constitution of the Republic of Uzbekistan], (Tashkent, Uzbekistan, 1992).

8. Kailash Satyarthi and Bupinder Zutshi, *Globalization, Development and Child Rights* (New Delhi: Shipra Publications, 2006), 71–72.

9. UNICEF, *The State of the World's Children 2004* (New York: UNICEF, 2004), 7.

10. See UNICEF, *First Call for Children: World Declaration and Plan of Action from the World Summit for Children* (New York: UNICEF, 1990).

11. Ibid., 6.

12. P. Anandharajakumar, *Female Child Labour* (New Delhi: A P H Publishing, 2004),13–14.

13. "Children under the age of fourteen years may not be employed or work in any public or private agricultural undertaking, or in any branch thereof, save outside the hours fixed for school attendance, the employment shall not be such as to prejudice their attendance at school." Article 1 in: ILO. 1921. C10: Minimum age (agriculture) convention. http://www.ilo.org/ilolex/cgi-lex/convde.pl.

14. June 26, 1973, 1015 U.N.T.S. 297, entered into force June 19, 1976 (Convention 138).

15. G.A. res. 44/25, annex, 44 U.N. GAOR Supp. (No. 49) at 167, U.N. Doc. A/44/49 (1989), entered into force Sept. 2, 1990.

16. UNICEF, *First Call for Children.*

17. 38 I.L.M. 1207 (1999), *entered into force* Nov. 19, 2000 (Convention 182).

18. Database of International Labour Standards. http://www.ilo.org/ilolex/cgi-lex/ratifce.

19. Elizabeth, D. Gibbons, Friedrich Huebler, and Edilberto Loaiza, *Child Labour, Education and the Principle of Non-Discrimination* (New York: UNICEF, 2005).

20. Confusion prevailed from the 1920s concerning the respective roles of the League of Nations and the UN in dealing with (child) slavery and of the ILO in dealing with child labor. This deficit regarding one of the worst forms of child labor persisted until 1999–2000.

21. C. K. Shukla and S. Ali, *Child Labour and the Law* (New Delhi: Sarup and Sons, 2006), 40.

22. Elisha Chanlett and G. M. Morier, *Declaration of the Rights of the Child, International Child Welfare Review* 22 (1968).

23. In 1976, the General Assembly proclaimed 1979 as the International Year of the Child (IYC). The general objectives were to promote the well-being of children, drawing attention to their special needs and encouraging national action, particularly for the least privileged and those at work. It was only during the IYC that international attention became fully refocused on the problem of child labor. The IYC acted as a spur to international agencies and civil society actors alike.

24. Alfred De Savza, ed., *Children in India* (New Delhi: OXFAM India, n.d.), 66.

25. Shukla and Ali, *Child Labour and the Law*, 41.

26. Alec Fyfe, *The Worldwide Movement Against Child Labour: Progress and Future Directions* (Geneva: International Labour Office, 2007), 8.

27. The idea of Wilfred Jenks was to consolidate the minimum age standards and to present a standard that developing countries would adopt.

28. Fyfe, *The Worldwide Movement Against Child Labour*, 10–11.

29. Kofi A. Annan, *We the Children* (New York: UNICEF, 2001), 80–82.

30. Cotton Campaign, http://www.cottoncompaign.org/frequently-asked-questions.

31. G. Tansibekova and V. Kosteckii, *Passport of the Young Citizen* (George Soros Foundation, 2004), 64.

32. See U.S. Embassy–Tashkent, unclassified telegram no. 3730, October 15, 2002. See Article 77 of the Labour Code as cited by U.S. Department of State, *Country Reports—2003: Uzbekistan*, Section 6d.

33. US Department of State, *Foreign Labour Trends Report: Uzbekistan 2006* http://www.state.gov.

34. US Department of State, Bureau of Democracy, Rights and Labour, *Country Reports on Human Rights Practices 2006: Uzbekistan*, http://www.state.gov/g/drl/rls/hrrpt/2006/78848.htm.

35. US Department of State, *Trafficking in Persons Report*, June 2007,http://www.state.gov.

36. US Embassy–Tashkent, unclassified telegram no. 2056, cited by US Department of Labor, Bureau of International Labour Affairs, http://www.dol.gov.

37. Ibid., n26

38. "Concluding Observations of the Committee on the Rights of the Child: Uzbekistan," Forty-second session, CRC/C/UZB/CO/2, 2 June 2006.

39. The ILRF requested copies of the submission and response from the ILO Committee in Geneva in August 2006 and was informed that these documents were not available to the public.

40. The mission was lead by ILO-IPEC and took place in June 2004. See ILO-IPEC, *CAR Capacity Building Project: Regional Program on the Worst Forms of Child Labour*, project document, RER/04/P54/USA (Geneva, September 2004), 1. The Government of Germany provided funding in 2003 to carry out these activities. ILO-IPEC Official, Active IPEC Projects as of May 1, 2004, USDOL Official, 2004.

41. Countries participating in the sub-regional project are Uzbekistan, Kazakhstan, Kyrgyzstan, and Tajikistan. See ILO-IPEC, *CAR Capacity Building Project*, vii.

42. Uzbekistan: Country Reports on Human Rights Practices 2003, United States Department of State (2004) http://www.state.gov.

43. "Further Growth in Uzbek Child Labour," Institute of War and Peace Reporting (June 18, 2004), available at http://www.iwpr.net.

44. Ibid., n6; also confidential reports on economic constraints in the cotton sector from Uzbek activists on file at ILRF.

45. EurasiaNet is funded by the Open Society Foundation under the auspices of its Central Eurasia Project.

46. Choihona/EurasiaNet, February 7, 2011, http://www.eurasianet.org/voices/choihona.

# FIVE

# Analytical Perspective

Cotton farming has expanded all over Uzbekistan including Kashkadarya, Samarkand, Bukhara, Khourizm, Sukhandarya, the autonomous Republic of Karakalpakistan, Namangan, Fergana, and Andijan. The main focus areas of our study were Namangan, Fergana, Kashkadarya, and Samarkand. We also found the opportunity to do case studies in Urgench and Khiva of the Khoerizmoblast as well as other areas of Uzbekistan that fall in the northwestern region of the country. These oblasts encompass much agricultural land and most of the population is engaged in farming.

The economic problems involved in the employment of children in the cotton cultivation of Uzbekistan also entail legal problems. The legal issues involved in the employment of children in cotton cultivation are: (1) minimum age (2) the fields in which they may be employed (3) the nature of their work (4) matters of leave and holidays (5) duration of work (6) and protection of health, safety, and welfare.

As there is an increasing tendency to view the institution of child labor from a welfare approach rather than a purely economic orientation, no study of this nature has so far been carried out in the cotton sector of Uzbekistan. The present study is basically an attempt to fill that gap. The significance of the present study has the following basis:

1. Children are the most valuable asset of a nation and deserve the best it has to give.
2. Fundamental rights and the directive principles of state policy enumerated in the constitution of Uzbekistan prohibit employment of children below age fifteen in any sort of work, protect children from abuse, forbid their exploitation, and ensure justice and humane conditions.
3. Children are engaged in various types of employment, but Uzbek children are mostly engaged in the cotton industry.

4. Children employed in cotton cultivation are badly exploited not only by the state machinery but by parents as well. Working conditions qualitatively as well as quantitatively are harmful to proper physical growth and mental, intellectual, emotional, and educational development.
5. Since Uzbekistan is known for its cotton throughout the world, demand in the local as well as outside market is very heavy. Because the state of Uzbekistan controls the industry, employers forcibly employ schoolchildren to meet production and price goals.
6. Though there is an abundance of research on child labor in general, no authoritative efforts have been made so far to study the schoolchildren working in the cotton industry in Uzbekistan.

The present study is concerned with issues, at both the individual and social levels. Among the important social issues, the life conditions of child laborers in different sectors of the economy deserve special mention because these children suffer in all aspects of life—physical, mental, moral, hygienic, and educational. The pathetic condition of these child laborers in present-day Uzbekistan has increased in the last two decades. In addition to their physical suffering they are economically exploited, which in turn degrades their ability to meet personal and domestic needs. Instead of attending school and enjoying their childhood, children are engaged in earning a livelihood, often in harsh circumstances. Economic compulsions weigh heavily on the poor parents, forcing them to collude with the child's employer in violating the law and putting the child under risk of human exploitation.

## REPUBLIC OF UZBEKISTAN

The sovereign and independent Republic of Uzbekistan, located in the middle of Central Asia, in the basin of the Amu Darya and Syr Darya rivers was formed in the wake of the collapse of the former Soviet Union. In the Soviet era, Uzbekistan functioned mainly as a raw resource base, had an unequal economic exchange, and had to follow the dictates of central authorities in Moscow. With the acquisition of independence, it became necessary to choose a form of government and a strategy for social and economic development that would ensure a viable transition to the world market. Uzbekistan's territory covers 447,400 square kilometers. Before it acquired independent statehood, the Uzbek Soviet Socialist Republic (Uzbek SSR) consisted of one autonomous republic (Karakalpakistan) and twelve oblasts: Tashkent, Syrdarya, Jizakh, Fergana, Namangan, Andijan, Kashkadarya, Sukhandarya, Samarkand, Bukhara, Khorezm, and Navoi.

The total length of the state boarder is 5,300 kilometers, a significant portion of which runs through deserts and plains. Only in the east does the boarder cross through valleys, canyons, and mountains (which are partially covered by snow and glaciers). Here Uzbekistan shares a border with Tajikistan and Kyrgyzstan. In the north the republic has an extremely lengthy border with Kazakhstan. In the west, the border mainly follows the *Amu Darya* River and sets the boundary with Turkmenistan. In the south, the republic shares a common border with Afghanistan. Two-thirds of the territory of Uzbekistan consists of desert, semi-desert areas, and steppe. The remainder is covered by fertile valleys. The climate in the north is moderate, yet also has sharp continental contrasts (with hot summers and snowless but cold winters); in the south, it is warmer, closer to fitting the criteria for subtropical. The level of precipitation is relatively low.

Uzbekistan has a well-developed agriculture sector, which accounts for 37.2 percent of the GDP (compared to 25.4 percent from industry, 10.4 percent from construction, and 27 percent from the service sector). Cotton is the main agricultural crop; Uzbekistan is the fourth-largest producer of cotton in the world. In 1991 it produced 4,646,000 tons of raw cotton, a decline of 422,000 tons from the previous year. But since 1995 the production of cotton has increased to a considerable extent and the production of 2010 was more than 8,000,000 tons. The sharp fall in the water level of the Aral Sea has led to noticeable changes in the volume and geography of participation and has also resulted in the formation of dust and salt storms.

The population of Uzbekistan has very deep historical roots and considerable ethnic complexity. These roots are many centuries old and define the integrity, unity, and interaction of the entire process of demographic and social development for the peoples of Central Asia. In 1991, the territory of Uzbekistan encompassed 20.7 million inhabitants, which jumped to 28.3 million in 2010, representing more than one hundred thirty different nations and nationalities. The largest ethnic groups included Uzbeks (70 percent), Russians (9 percent), Kazakhs (4 percent), and Karakalpaks (2 percent).[1] The settled way of life for the native population and the high level of agriculture have historically predetermined the dominant proportion of the rural population in Uzbekistan (60 percent), a low level of mobility, and a strong attachment to home areas. It is no accident that a popular saying holds: "If you lose your family or kin, you will weep for seven years. If you lose your native land, you will weep for your entire life."

From ancient times, people in the east have believed that children make a family strong. Therefore the Uzbeks (by religious confession), like all Muslim peoples, believe it a pious deed to have as many children as Allah sends. Hence, in Uzbekistan, for example, in 1992 the natural growth rate of the population (per 1, 000 inhabitants) was 26.6 (compared to 26.9 in Turkmenistan, 25.6 in Tajikistan, 21.4 in Kyrgyzstan, and the lower rate of 11.8 in Kazakhstan). These indices are striking when compared with data for the European countries—for example, 2.2 in the United Kingdom, 0.9 in Germany, and 3.8 in France. According to data from the Central Statistical Administration of the former Uzbek SSR, the size of the average family in Uzbekistan was 5.5 individuals (compared to 3.2 in the European republics of the former Soviet Union). The high birth rate has meant a significant increase in the population of Uzbekistan, which now ranks only behind Russia and Ukraine among the states of the former Soviet Union. This demographic characteristic of Uzbekistan has also affected the social structure: 43.1 percent of the population is under the age of fifteen (compared to 24 percent in Russia and Belarus, and 33.2 percent in Kazakhstan).

The economy of Uzbekistan was shaped and developed to satisfy, first and foremost, the "all union" interests of the USSR as a whole. This meant a multifaceted exploitation of its minerals and raw materials, energy, land, water, labor, and other resources. The general economic plan of the Soviet Union predetermined the place of Uzbekistan in the so-called "all-union market." Concretely, this meant specialization in the production of materials that the Soviet Union deemed strategically important. These included agricultural products (raw cotton, the cocoons of silkworms, ambary stalks, fruits, vegetables, karakul or Astrakhan lambskin, wool, etc.) and industrial goods (wool fibers, threads of raw silk, mineral fertilizers, machinery and equipment for the preliminary processing of products from the cotton industry, and semi-finished goods from gold mining and chemical industries). The struggle to make the Soviet Union self-sufficient in the production of cotton proved an onerous experience for the Uzbek people, for it resulted in the creation of a one-sided agricultural system. In 1990, for example, cotton represented 54 percent of the total area of sown acreage, a proportion that was still higher in certain areas (e.g., 68 percent in Andijan oblast and 60 percent in Bukhara oblast).

As a result of Moscow's predatory policy, cotton production dominated all other sectors, with industry itself being dedicated primarily to servicing the cotton complex. Moreover, all sectors were oriented toward providing the Soviet market with such goods as semi-processed chemicals, cotton fiber, semi-processed copper and gold, and the like.

## SAMPLING

In the present study, multistage sampling was used. In the first stage Tashkent, Kashkadarya, Samarkand, Namangan, Khiva, Urgench, and Fergana were selected because of (1) more land under cotton cultivation, (2) employment of a large number of children in cotton cultivation, and (3) easy accessibility. Non-probability sampling technique was used for the present study. In many research situations including the present one and also those where there is no list of persons to be studied (e.g., alcoholics, wife battering, migrant workers, etc.), probability sampling is difficult and inappropriate to use. In such researches, non-probability sampling is the most appropriate one. Keeping in view the above criterion we found the non-probability sampling the most appropriate to get the desired results. Non-probability sampling procedures do not employ the rules of probability theory, do not claim representativeness, and are usually used for qualitative exploratory analysis. The five types of non-probability sampling that were used in the present study for the collection of data from the respondents are: convenience, purposive, quota, snowball, and volunteer.

The present study followed the qualitative methods of data collection, based on non-standardized interviews with open-ended questions. We interviewed and interacted with persons comprising schoolchildren, parents, employers, teachers, farmers, university professors, social workers, and other local people. In addition to the interviews we also collected data from different state and non-state organizations, which include especially university libraries, agricultural institutes, economic faculties, cotton institutes, state libraries, UNICEF, UNESCO, UNDP, and World Bank Offices of Uzbekistan, and various other organizations and departments.

## ANALYTICAL APPROACH

The practice of child labor in Uzbekistan represents a distinctive case. Global patterns suggest that it is generally families and/or employers who tend to be the major initiators and beneficiaries of children's work. Uzbekistan represents a rare instance of state-sanctioned mass mobilization of children's labor. The principal beneficiaries are not households or primary producers but state-controlled trading companies higher up the value chain. Their exclusive control over cotton export revenues and their ability to appropriate the profits generated by the disparity between domestic and international market prices gives them a stake in the maintenance of the current procurement and labor-control regime.

During the course of field work, we observed some striking facts about child labor in the cotton industry of Uzbekistan. Our empirical findings brought out some revealing facts about the institution of child labor in Uzbekistan. We witnessed some special cases of child laborers that highlighted the nature, causes, and consequences of the problem. The study of these cases will certainly lead to a deeper and wider sociological understanding of the phenomenon of child labor in the cotton industry of Uzbekistan.

## EFFECT ON THE EDUCATIONAL SYSTEM

Since independence, Uzbekistan's children have been less likely to go to school than they had been under Soviet rule, when education was a priority. Today budgets and political priority have left the schools with outdated and irrelevant textbooks, untrained teachers, and crumbling school buildings. The government's policy of requiring students to pick cotton for three months a year has contributed to a lack of respect for education. Since independence, the literacy rate (the percentage of people older than fifteen who can read and write) has declined from more than 99 percent to less than 97 percent. As schools continue to close, especially in the rural areas, the literacy rate is expected to fall even lower. Schools often close for the months of September, October, and November so that children can pick cotton and educational facilities can be used to house cotton pickers brought in from other areas. Even the youngest children, seven or eight years old, go to the fields. Those who do not meet their quotas or who pick poor-quality cotton must serve detention and receive reduced grades, which lower their prospects for employment or further education. Other punishments include scrubbing floors or fetching drinking water from great distances. Those who meet their quotas receive 2 to 3 cents for each kilogram of cotton they pick.

Dilnora, around fifty years old, in the presence of her teenage daughter, talked to us about the cotton cultivation in Uzbekistan.[2] She narrated her experience and observation that Economic and Agriculture University students are sent into fields to pick cotton in order to meet state cotton quotas. She also said that this time only boys were sent to pick cotton. Lessons were closed from September onward while cotton was picked from the field. Each student had a quota of thirty kilograms to be picked in one day and received 20 soms for picking one kilogram of cotton and in this way they earn 600 soms for picking thirty kilograms in a day. If any students do not want to pick cotton, they have to pay a huge amount ($1,020) according to Dilnora, an unbearable sum, so it is better for them to be in the field than to be in class. They are sent to fields with the justification that it is the practical part of their study to be in the cotton field, as they are students of an agricultural and economic university.

Abdul Sattar, a teacher from one of the schools in Kashkadarya said that the first bell rang in schools in the first week of September in Uzbekistan this year.[3] "The 'first bell' in Kashkadarya schools not only means that children for the new academic session are back in schools, but also that the cotton harvest season has begun for schoolchildren and teachers." Sattar divulged that two and a half months are designated as the entire academic term and the first term of the school year is spent picking cotton. He also said that teachers register pupils who are picking cotton as being present in class and even mark them with better grades if they pick cotton really well. This year the cotton harvest in Uzbekistan was and is mostly done by students from schools and colleges. According to Sattar, the director of the school instructed teachers to prepare schoolchildren for cotton picking. The school teacher Sattar reported that even in the first day of the new academic session in school he had to prepare children to pick cotton. "We had also to take them to the cotton fields. Poor kids wear nice, new clothes and within only a couple of days they are made to pick cotton with the same clothes," said the teacher from Kashkadarya.

Sattar says that children pick cotton in fields close to the school and teachers are sent in groups to districts farther away. "Up to now, the only news is that the list of teachers is made and starting from September 3, an average of fifteen teachers from each school will pick cotton for fifteen days. We are lodged in a hut in the farm field. We were there last year during the harvest. The food and water is no good. We put a blanket on concrete and sleep there. There is no drinking water. The water is salty. It is uncomfortable to bathe there. The conditions are very bad. There is no choice; we go for fifteen days and tolerate it. Cotton picking is state policy." Sattar also revealed that for a teacher both the classes and the cotton are part of the job.

## EXPLOITATION OF FARMERS

Bekmuratov,[4] a farmer from Bukhara, says that Uzbek farmers do not benefit from the rise in cotton prices on the world market because in Uzbekistan the government buys the cotton from farmers for a price they like. In Uzbekistan, the difference between the cotton production expenses and the price the government pays is very little; in other words, a farmer nets only about 10 to 15 percent from the entire crop, he says. "If the harvest is good and if we meet our contracted target, the profit will be 10 percent. If let's say we spent 100 million soms on our expenses, then our profit will be 10 million. Expenditure is up to 85–90 percent," says Bekmuratov.

Karimov, another farmer, says that the main profit from cotton goes to the government. "The government buys the first sort cotton for 640,000 soms from us and the second sort for 600,000 soms. The farmers get only 50 percent of the real price of cotton." The rest goes to the authorities, says Karimov. "In these days farmers do not benefit from growing cotton; when they harvest the crop in autumn, the selling price and expenditure incurred is almost equal. The price of mineral fertilizers is on average 300,000 soms per ton. One liter of diesel is 1,200 soms. A tractor costs 27 million soms. You can calculate the cost of amortization. The state taxes are so heavy. Farmers pay sixteen different taxes. We pay 450 per 1,000 soms that we take for wages." According to Bekmuratov, as long as the

Uzbek farmer is not independent from the state, no matter how much the cotton price rises on the world market, the farmer can never gain. Every farmer is controlled; internal affairs, prosecutor, and all other structures of the state control the farmers. As if there is no other problem in the society except farmers. There is only one thing—live or die but you have to meet the planned target," says Bekmuratov.

## HUMAN RIGHTS

A human rights activist interviewed claimed that Uzbekistan has a cartel running cotton production and export.[5] Uzbekistan pledges more or less 60 percent of the state's hard-exchange export earnings. On the other hand, the farmers are remunerated very shoddily for their work; they are mostly dependent on quotas and are forced to grow a certain amount of cotton and to sell it for unrealistic, low prices to the government. Ninety percent of the Uzbek cotton is still handpicked. To guarantee the export incomes through low investment, the Uzbek government organizes a cotton campaign mobilizing schoolchildren, students, and people working in the public sector to work on the fields every year. All profits from the cotton sector are concentrated in the hands of the president, his family, and those few close to him. The activist divulged that the mobilization for the cotton harvest is state sanctioned and is implemented through a strict hierarchic system of pressure and intimidation on all levels. Depending on the quotas imposed on the region, the local authorities (hokimiyat) decide how much cotton the schools and the farmers of the region should pick. The farmers and school administration agree on the scope of the work, passing the information on to teachers and schoolchildren.

Farmers in most cases are in charge of the transportation of the pupils to the fields as well as their meals. Schoolchildren who do not want to participate in the cotton harvest are forced to do so by the school administration and teachers, who threaten bad grades as well as direct humiliation through publicly insult and rebuke. Sometimes when the parents do not allow their kids to work on the fields, the local authorities threaten them with dismissal or even a criminal charge.

The human rights activist also disclosed that the schoolchildren have to work on the fields for about eight hours a day, every day for two months (from mid-September to mid-November). Children scarcely receive food from the farmers and thus bring their food themselves. The cotton fields are often sprayed with defoliants, which additionally affects the health of children and young people working there. The daily quota, which schoolchildren are obliged to fulfill, lies between thirty and fifty kilograms, which is paid by 100 UZ soms (.07 USD) per kilo. In addition, the schoolchildren are supposed to spend a part of this small remuneration for transportation, food, and sometimes even a contribution to the school.

The human rights activist went on to say that Uzbekistan and Turkmenistan are among the world's most repressive states, with no free elections. Opposition activists and human rights defenders are subject to persecution. The lack of a free media allows many abuses to go unreported. Unelected local governments are usually complicit in abuses, since they have little or no accountability to the population. Cotton producers have an interest in continuing these corrupt and nondemocratic regimes because the industry relies on cheap labor. Schoolchildren are still regularly required to spend up to two months in the cotton fields in Uzbekistan. Despite official denials, child labor is still in use in Tajikistan and Turkmenistan. Students in all three countries have to miss their classes to pick cotton. Little attention is paid to the conditions in which children and students work, which results in bad health and even some casualties.

## STATE RESPONSE

According to the social worker Shaharzad,[6] all the schools and colleges are closed and year-one children stay at home while teachers and other schoolchildren work in the fields." He also mentioned that the security presence in the fields has increased.

Shaharzad also said that owing to good weather conditions, this year's harvest was a good one and the authorities have set a target of over 3.5 million tons of raw cotton. To achieve this they will need to recruit more "volunteers" than ever. He also said that in the central region of Jizak, markets in a number of districts remained closed for the peak season of cotton cultivation. "Everyone is picking cotton," he explained. Coercion is overlaid with propaganda, as the government appeals to the nation's sense of patriotism. On September 15, *Khorezmskaya Pravda,* a state newspaper in northern Uzbekistan, published an appeal headlined "Cotton Is Our Pride, Our National Wealth." "Every one of us should be working in the fields today," the statement said. It is believed that these advertized programs in newspapers and local television channels of motivating and sensitizing people toward cotton picking are completely state backed and planned with the intention of meeting the set quotas of cotton. Muslim clerics in the eastern Andijan region have been drafted into supporting the campaign. An article in a local newspaper said "picking every gram of cotton is a sacred duty for every Muslim." The social worker observed that in various regions the authorities are using various kinds of rhetoric to obscure their interest in forced labor. Some education division employees in various parts of the country including Tashkent say that there is no issue of compulsion, but Shaharzad says they are justifying it with the argument that we have been picking cotton for twenty years and our children will gather it also.

## COTTON AND ENVIRONMENTAL POLLUTION

The Uzbek cotton industry is a ruinous eccentricity created by Soviet central planning.[7] Over 80 percent of the loss of water from the Aral Sea is due to irrigation for the Uzbek cotton industry, so it is one of the World's greatest environmental disasters. On most agricultural land in Uzbekistan, cotton has been grown as a monoculture for fifty-five years, with no rotation. This of course exhausts the soil and encourages pests. As a result the cotton industry employs massive quantities of pesticide and fertilizer. The Aral Sea is disappearing, suffering appalling pollution, reflected in abysmal levels of disease. The professor disclosed that Uzbek farm workers are tied to the farm. They need a *propusk* (visa) to move away—which they won't get. The state farm worker normally gets two dollars a month. Their living and nutritional standards would improve greatly if, rather than grow cotton, they had a little area to grow subsistence crops. The professor said that there are no independent research institutes allowed in Uzbekistan. In fact the proportion of the population enslaved on state cotton farms is closer to 60 percent. This system is only sustainable under conditions of political repression, which can be used to mobilize workers at less than market cost.

Sator, a professor in the university, told us that the condition of state-forced child labor in the Uzbek cotton fields is not sensationalist; they are very much the everyday conditions in which hundreds of thousands of Uzbek children are forced to live for months.[8] Women do much of the hard manual labor in the cotton fields, and reap almost none of the benefits. Cash wages are minimal and often paid late or not at all. In most cotton-producing areas, growers are among the poorest elements in society. The environmental costs of the monoculture have been devastating. The depletion of the Aral Sea is the result of intensive irrigation. The region around the sea has appalling public health and ecological problems. Even farther upstream, increased salinization and desertification of the land have a major impact on the environment. Disputes over water usage cause tension among Central Asian states.

## COTTON TRADE

The executive summary from the International Crisis Group report on Central Asian cotton says[9] :

> Central Asian cotton is traded internationally by major European and U.S. corporations; its production is financed by Western banks, and the

final product ends up in well-known outlets in Western countries. But neither the international cotton trading companies nor the clothing manufacturers pay much attention to the conditions in which the cotton is produced. Nor have international organizations or IFIs done much to address the abuses. U.S. and E.U. subsidy regimes for their own farmers make long-term change more difficult by depressing world prices.
To effect immediate change, you should demand that your apparel manufacturer state on their garment labels where their cotton comes from, and that it does not come from Uzbekistan. With the vast volume of T-shirts bought and sold, the message will quickly spread, and High Street retail will follow. Why am I doing this? As a large user of cotton, and with our influential position in the T-shirt industry, Continental Clothing has an opportunity, if not even a responsibility, to raise awareness and promote consumer action on issues where we feel strongly—such as the state orchestrated child slavery in Uzbekistan. The wonderful thing is that it costs us nothing, and may cause consumers to question the garments they buy and so switch them on to cotton garments which guarantee that certain positive social and environmental conditions are met—such as Continental garments. This is often the way with ethical and environmental choices, initially they appear expensive and difficult, until you realise they can be sustainable choices for a longer term and more profitable future. So yes, we are doing this because we can, and also for personal gain. If you follow the same formula, you may benefit in exactly the same way.

## IMPACT ON DOMESTIC LIFE AND WELL-BEING

"I am the mother of six children.[10] Already around one and a half years ago my husband and eldest son went off to work in Russia. I married off two of my daughters. Two of my sons study in the district center, one in the seventh grade, the other in the ninth. This year was the first year they forced my seventh grader to go out to pick cotton. He is a scrawny boy and two years ago he got ill with hepatitis. Next to our residence we have a garden plot, where we produce fruits and vegetables to vend and somehow make a living. This year the chairman of the collective farm insisted that I go out to pick cotton and my daughter in law and remaining children; otherwise, he would take our plot away. How can I go out to pick cotton?" Desperation on her face, tears started to flow down from her eyes. "My daughter-in-law is pregnant. The chairman said that if we don't go out, I'll have to pay one hundred thousand soms [approximately 60 dollars, or more than two average monthly wages].When I said there was no means I could pay that kind of money, he said that in that case we wouldn't obtain the benefits from the *mahalla*."

Matters can turn out to be even worse when parents who keep their children absent from the harvest may be charged with an offense.[11] A tenant of Namangan of Village Giganth reported his fairy tale like this: "Don't even inquire about that, I am not going to respond to the query. I have breathed in this village for thirty five years. All my life, and that of my wife, has been spent out in the fields. We have three children. They go to the school across the street. The oldest boy is in the eighth grade. On 5 September their teacher told them that they are going out to pick cotton. My son and some five other boys objected, and said they would not go. The teacher let the other students go, but kept back my son and the other boys in the classroom and beat them up. My son came home in the evening in tears, with a bloated face and two black eyes. The next day I went to the school and met with the course director. He refused to take note and called me an enemy of the people. I said to him that I have a crooked back from being in these cotton fields my entire life and enough is enough. My son is going to breathe in a different way. I want him to study, to become knowledgeable. The course director started to bash his fists on the desk and say that he would call the police, that I must stop haranguing him, and threw me out. He spoke to me in a very indecent language, and truly snubbed me. So even though we didn't have any additional food in the house, I packed whatever was available for my children's meal to be out in the fields. What else could I have done? After all, the solitary thing they provide to the cotton pickers for lunch is several potatoes, cabbage, or macaroni. Previous year five or six schoolchildren, from my son's school broke into houses simply to pilfer some food. it took their parents a long time purely to get them out of the police station. So, those were my feelings as I left the school."

The farmers who make use of the children's labor often admit that they have no time or resources to devote to the children's health.[12] The farm director recounted that "each year when the schoolchildren come to pick cotton in my fields, I myself supply them with drinking water. Using special barrels for that purpose from the garage of the farming association, I bring drinking water from the taps near the district center. In a few cases when there was no water in the pipes, I brought water from the irrigation canal that flows through the district. It's hard to say anything about the quality of that water. After all, I grew up on that water myself. And I am healthy. However, the doctors from the Central Hospital are always saying that you at least need to chlorinate this water. But not everyone follows this recommendation, because we are sure that this water wouldn't poison anyone. A few doctors have said that from pollution or because the water itself is of low quality one can get kidney stones, or enlarged spleen or liver. As a farmer, it is my responsibility to fulfill the production quota and productivity of the harvest. Let their parents worry about the health of children."

"For the last four years I have been coming out for cotton picking.[13] We have to buy the school uniforms with the money I bring home from the cotton harvest. So it is a way I can help my parents. Last year right after the cotton campaign, I got sick with hepatitis. The doctor said I got infected from dirty water. The tractor that was supposed to bring us drinking water broke down, and we had to drink from the irrigation canal. Along the borders of the canal they had spread saltpeter and so many kids got poisoned. Some of them got sick like I did. However no one gave us any medical assistance, or medicine. When the chairman of the farm, the district *khokim*, and the auxiliary policeman came out into the cotton fields, they threw stones at the kids who weren't picking because they were sick. And if any of them got hold of you, they would beat you. A few students, being afraid of them, would run from field to field and hide."

"It is a clear and open fact here that often children are not paid the full amount due to them.[14] Their payment depends in part on the grade of cotton they pick, which is itself tied to the various stages that the harvest entails. At the start of the cotton harvest season the government sets the price for one harvested kilogram of 50 soms (3 US cents). However by the middle of the season the local administration lowered this to 40 soms, and by the end of the season, to 30. If on average a pupil receives 40 soms per kilogram, then on average he or she earns 1,600 soms (less than one US dollar). Over two months of harvesting that amount may grow to about 96,000 soms (60 US dollars). If one takes into account that students spend about 1,300 soms per day just on food, it follows that they are working in the fields practically for free. For instance, all of the textbooks required for seventh grade cost approximately 25,000 soms. A student will not even be able to buy their schoolbooks with his hard earned money for months of labor in the fields. In this sense the students are truly slaves."

"My daughter is in the ninth grade and my son in the seventh.[15] Every year, both of them are sent by their schools out to pick cotton. Naturally, we don't object, because at least they can bring in a little money to help with the cost of their clothing, schoolbooks and supplies. However, I do object to sending girls who are close to being of sensitive age. After all, girls are not so suited to hard physical labor and to those terrible living conditions. Even if I allow my daughter to go out and pick cotton, she earns 100,000 soms, but when I marry her off I will have to pay 500,000 just to treat her various ailments. But my husband has left to work in Russia, and that's why I agreed to let my daughter go. Otherwise, I would have gotten a medical certificate from the district hospital. How can a mother allow her young daughter to be out in the open and work; it is only the pressure of the state agencies which are putting the young girls in the fields for cotton picking. Girls are not as sound as boys from the physiological and biological point of view and are susceptible to more bodily damage, like bruises and stresses on the sensitive parts, than boys. So it is a matter of shame on our state which puts the innocent young girls into cotton picking who are the mothers of tomorrow and also being Uzbek and Muslims they should not be doing it.

"After all, children are children.[16] You cannot shield them completely from jeopardy. They need constant care and are to be monitored consistently. Throughout the cotton collection season I was able to learn by heart cases where children were asphyxiated by the cotton they gathered, or were scuttled over by tractors when they were sleeping in the fields. Sorry to say it that due to the national disposition, whether it is out of heedlessness, disproportionate good-heartedness, or trepidation, persons answerable for these catastrophes are time and again exonerated. Both national as well as international press as well independent ones doesn't put pen to paper on the subject of these occurrences, nor are they revealed on small screens. For instance, I am acquainted with a case when six adolescent girls who walked off to wash in a river or some canal were carried away by the current. The prosecutor's office even lodged a scandalous case; however, afterward they put the brake pedal on it and it never went anywhere.

## NOTES

1. There are 5.5 inhabitants per square kilometer in Kazakhstan, 10 in Turkmenistan, 28 in Kyrgyzstan, 49 in Tajikistan, and 64 in Uzbekistan; detailed annual figures available athttp://perspective.usherbrooke.ca.
2. Interview with a housewife from Tashkent. The interviewee was aware that his/her remarks might be published in a book/journal.
3. Interview with a school teacher in Kashkadarya (the heart of Uzbek cotton). The interviewee was aware that his/her remarks might be published in a book/journal.
4. Interview with a farmer in Bukhara. The interviewee was aware that his/her remarks might be published in a book/journal.
5. Interview with a human rights activist. The interviewee was aware that his/her remarks might be published in a book/journal.
6. Interview with a social worker in Samarkand. The interviewee was aware that his/her remarks might be published in a book/journal.
7. Interview with a professor from Tashkent. The interviewee was aware that his/her remarks might be published in a book/journal.
8. Interview with a professor in Fergana. The interviewee was aware that his/her remarks might be published in a book/journal.
9. Available at https://octoberonline.wordpress.com/tag/central-asian-cotton/.
10. Interview with a woman from Namangan. The interviewee was aware that her remarks might be published in a book/journal.
11. Interview with a farmer in Namangan. The Interviewee was aware that his/her remarks might be published in a book/journal.
12. Interview with a farm director in Namangan Oblast. The interviewee was aware that his/her remarks might be published in a book/journal.
13. Interview with a schoolchild from Namangan. The interviewee was aware that his/her remarks might be published in a book/journal.
14. Interview with a school teacher from Namangan. The interviewee was aware that his/her remarks might be published in a book/journal.
15. Interview with a mother from Fergana. The interviewee was aware that her remarks might be published in a book/journal.

16. Interview with a school teacher in Fergana. The interviewee was aware that his/her remarks might be published in a book/journal.

# SIX

# Sociological Perspective

Uzbekistan's largest economic sector is agriculture, and cotton is its main crop. Around 75 percent of the cotton is exported, making Uzbekistan the second-largest exporter of cotton in the world. The cotton is grown on government-owned plantations where workers are urged by government newspapers to "reap prosperity." Farmers are allowed to lease parts of the government plantations, but they can only plant cotton, and they are told how much they are expected to produce. Failure to meet their goals can cost them their farms.

As an export crop, cotton generates most of Uzbekistan's income from outside the country and thus also makes up most of its annual revenues. Since independence, Uzbekistan's government has raised the quotas for cotton production each year, though they are not often met. The quotas have resulted in systematic corruption. Farmers seeking to avoid penalties for not yielding a prescribed amount of cotton, pay huge bribes to local officials, who falsify the production numbers. As a cash crop, cotton has also become less profitable each year. The world price for cotton and the demand for the fiber have both dropped, as has Uzbekistan's ability to meet its production goals.

Since their days as Soviet citizens, most Uzbeks have been obliged to devote the autumn months to picking cotton. September is harvest time, during which thousands of workers head to the fields. Schools close until December and other industries shut down as well. Children as young as seven work in the fields, supervised by teachers who are held accountable for their students' production. For up to three months each year, the children are housed in shabby barracks, farm sheds or school rooms. They live on diets of macaroni, bread, and sweat tea made with untreated water. The amount of food they receive depends on how much cotton they pick. Medical treatment is either unavailable or denied, so children return to school sick or malnourished, unable to make up the work they have missed.

Uzbekistan refuses to sign the international convention agreements that prohibit child labor, saying the country does not force anyone to pick cotton. Leaders claim that everyone volunteers out of love for their homeland, which is absolutely wrong. Wealthy families will sometimes bribe harvest officials to obtain a certificate of poor health for their children, thus exempting them from the harvest. But it is an option that is open to few. The cotton crop has exacted a toll on Uzbekistan's older students as well. University students in Bukhara spend two months in the cotton fields at harvest time, where they are required to pick 176 pounds (80 kg) a day and reimburse the government for their food and lodging. Officially, the four thousand students are volunteers so they are not paid. In the Kashkadarya province, near Kashka River, hundreds of thousands of workers are forced to harvest cotton under a system that dates back to Soviet days. In Kashkadarya, the heart of the nation's cotton industry, the harvest is sometimes as large as four hundred twenty thousand tons (380,940 metric tons). Workers are paid 15 to 20 Uzbek soms, or 1 to 2 cents, for each kilogram of cotton they pick. From this they pay for their food and lodging in run-down barracks. The government does not reveal its selling price for the cotton because economic figures are often classified as state secrets.

Future plans for the cotton industry include improvements in the textile manufacturing sector, where machinery is seriously outdated, badly damaged, or simply worn out. The government hopes by improving the textile industry it will be able to employ more workers and to use more of the cotton within the country.

A group discussion is very helpful method of getting good amount of information. The group interviews we conducted in different provinces of Uzbekistan used a systematic order of questions so that a sequence in investigating the problem could be followed. The group discussion highlights and bring into the picture all the elements and components that are part of this inhumane institution existing in Uzbekistan at the behest of state. People belonging to different echelons of society participated in the discussions. In the following paragraphs each variable is discussed in detail.

## LEVEL OF FORCED CHILD LABOR IN UZBEKISTAN

Every year in the autumn all schools are closed to have children aged eight to fifteen in Uzbekistan harvest cotton. The children enrolled in schools in the capital and other major cities are not involved, as urban areas are monitored by social activists, human rights groups, and local and non-local media. Children studying in lower grades in lyceums and college and university students, as well as local public servants are also necessary to labor in cotton fields for the period of the harvest. Children in rural regions are also by and large required to weed the cotton fields in the spring. In entirety, schoolchildren neglect three to four months of lessons each year.

According to appraisal and coarse reckonings, child labor accounts for more or less half of the entire cotton harvest in Uzbekistan. Cotton pickers, labor seven days a week up to ten hours a day with a solitary short break for lunch. Children who do not reside close to the harvest camp in the cotton fields stay in living environments devoid of hygiene and other basic amenities of life. A good number of them suffer malnutrition and lack medical services. Every day children are given a quota of between thirty and fifty kilograms of cotton. If their quota is not met, children are deprecated by the school administration, and sometimes physically maltreated.

## INSTITUTING CHILD LABOR IN UZBEKISTAN

The use of child labor in Uzbekistan is a calculated state policy. Instructions transferring children to the cotton fields are issued by school commissioners who receive instructions from the home administration (*hokimiyats*). Given Uzbekistan's austerely centralized system of power and control and cotton business management, it can be construed that the guidelines come from the central government.

## THE APPARATUS OF HOUNDING

Directorial castigation coerces children, as well as lyceum, college, and university students, to labor in the fields. The children and students are threatened with eviction if they do not discharge their responsibility to pick cotton. Parents who do not send children to the cotton collection risk losing public welfare reimbursements. School supervisors and even instructors can lose their posts if they fail to act in accordance with the instructions of local authorities. Throughout the cotton harvesting, Uzbekistan becomes a virtual labor encampment with children and tutors organized in labor contingents patrolled by a police force.

## EXISTING LEGAL PROTECTION

Uzbekistan has adopted a number of laws, on youth policy, labor, and children's rights. These decrees put limits and restrictions on the use of child labor. Lawfully, children are only permitted to work after age of fifteen. Prior this age they can be engaged only in the framework of school-related activity, and for no more than fifteen days. In every case, children are not permissible to work more than four hours per day. These legal provisions are not imposed, and the use of forced child labor is common all over Uzbekistan.

## PROFITS FROM THE ESTABLISHMENT OF ENFORCED CHILD LABOR

This dominant agricultural industry of Uzbekistan is stringently administered by the central government in Tashkent and all procurement prices of cotton are decided and managed from Tashkent. Cotton and grain planting uses two-thirds of the best arable lands in Uzbekistan. Cultivators have neither the right to decide which crops to sow, nor to whom they vend their yield. As in Soviet times, the Uzbek government imposes cotton production quotas on all farmers and local administrations. The local *hokims* (governors) are personally answerable for these quotas. Farmers cannot sell cotton in the open bazaar at market value and must bring their harvest to local government cotton gins. Farmers endeavoring to sell their produce to neighboring countries are prosecuted for smuggling and face punishment with fines.

Cotton is the principal export article of trade and key resource of revenue for the kingdom. While three trading corporations formed at the Ministry of Foreign Economic Affairs trade cotton on global markets, all export returns remain under the control of executives employed by President Islam Karimov. The government of Uzbekistan provides no information on the subject of export revenues or the value of Uzbek cotton vended in worldwide markets. Even less is known regarding how cotton revenues are distributed within Uzbekistan, although it is implicit that substantial sums are funneled to the bank accounts of the president and other high bureaucrats.

## PRIMARY CAUSE OF CHILD LABOR

The most important reason for the establishment of enforced child labor is the domineering state control. The Uzbek government enforces unjustly low procurement prices on cotton, then sells the cotton at market prices. Thus, cotton farmers and workers engaged in the cotton sector are increasingly impoverished and resort to relocating to neighboring Kazakhstan and Kyrgyzstan, where they can make respectable earnings. Given these circumstances, the Uzbek government prefers using forced labor to implementing reforms. It relies on intimidation, with the use of police and legal procedures against farmers who do not accomplish their anticipated cotton and grain proportions.

## CHILD LABOR AND SOVIET LEGACY

During the Soviet epoch Uzbekistan achieved a relatively high level of social and economic development, even though these attainments occurred in the midst of a number of severe social and environmental issues. About half of the cotton used to be collected by machines. Nowadays, as a consequence of corruption, lack of restructuring, botched incentive systems, and unbalanced allocation of cotton revenues, the use of machinery is severely reduced. Currently, in spite of various minor developments (for example, the beginning of quality control in wrapping and stocking systems), the cotton industry as a whole is returning to primitive conditions. The degree of forced labor has correspondingly grown as mechanization has decline. Collapsing social and economic conditions associated with the deterioration of the cotton sector have been particularly destructive in rural areas.

## WESTERN RESPONSE

The reaction from the West has so far been unhelpful. The United States and European Union protect Uzbek cotton and textiles under the Generalized System of Preferences (GSP), which gives import tax exemptions to underprivileged developing countries. In Europe, Uzbek cotton was approved a GSP rank in June 2005, immediately following the Andijan mass execution in which Uzbek security forces fired upon unarmed protestors killing hundreds. The GSP was proposed to deal with discrepancies in business dealings sandwiched between North and South, however in the case of Uzbekistan, it tends to reinforce social injustices. GSP allows the autocratic Uzbek regime to grow more affluent while the systemic abuse of child labor is perpetuated.

## COUNTRIES AND COMPANIES CURRENTLY IMPORTING UZBEK COTTON

Uzbekistan has moved their cotton exports from European to Asian countries for the past fifteen years, principally China and Bangladesh. However, after being processed in Asia in textile and clothing factories, Uzbek cotton makes its way into Western markets. Several Western countries also continue to buy Uzbek cotton directly. Various European countries, primarily the Czech Republic, Austria, Poland, Greece, and Italy buy Uzbek cotton indirectly. Coincidentally or not, these very same countries of late opposed extension of EU sanctions against Uzbekistan that were adopted in reaction to the Andijan mass murder in May 2005. Companies mainly active in importing Uzbek cotton purportedly include: Cargill Cotton, Weil Brothers and Stern, Plexus Cotton (UK), Daewoo

Textile Company (Korea), and Indutech (Italy), and others. There are also certain banks that finance and help companies to base their operations on and provide financial support to importers of Uzbek cotton.

## INTERNATIONAL INSTITUTIONS AND GOVERNMENT AWARENESS

Forced child labor has been prevalent in Uzbekistan since the Soviet times, but it became known to the global community only when Uzbek human rights activists and journalists decided to break the conspiracy of silence. The lack of international knowledge of the state of affairs in Uzbekistan provides the opportunity to continue exploiting young schoolchildren. The International Cotton Advisory Committee was forced to change its position on the issue from ignoring it to acknowledgment, when its leaders met with Uzbek activists and experts. The European Union is still ignoring the subject, thus allowing the Uzbek regime to remain exploiting schoolchildren without critical scrutiny.

## WEST'S RESPONSIBILITY

Companies and governments should urge the government of Uzbekistan to put an end to the use of compulsory child labor in the cotton industry. Until it is verified that this practice has ceased, Western companies, cotton importers, and retailers should refrain from purchasing Uzbek cotton and textile products. Moreover, Western governments must revoke GSP standing from Uzbek cotton imports and textile products. This trade benefit may be reestablished when it is verified that mandatory child labor practices have ceased. To reject labor abuse of children in developing countries, the EU and other Western countries must establish a certification and cataloging system that would allow retailers and consumers to scrutinize the source of cotton products imported and put up for sale. The International Labor Organization (ILO) and UNICEF ought to, albeit belatedly, investigate and publicize the issue of forced child labor in Uzbekistan in order to bring much needed international attention to this injustice.

## BOYCOTT OF UZBEK COTTON

The true producers and pickers of cotton in Uzbekistan get a trifling share of overall cotton export returns. The adult rural population of Uzbekistan is consequently boycotting Uzbek cotton by leaving the country to search for fair prices and sustainable earnings. Though child laborers get some return for picking cotton, it is small and comes at an objectionable price, the harm to their education. A boycott of Uzbek cotton would compel the Uzbek government to take genuine measures for restructuring its cotton industry, as a consequence of which Uzbek farmers would receive returns analogous to cotton farmers in neighboring Kazakhstan, where the government does not meddle in the sector.

## RATIFICATION OF ILO CONVENTIONS

Though the government of Uzbekistan ratified both the conventions (138 and 182) of the ILO, the guidelines of these conventions are never put into practice. Despite the fact that the ratifying of the conventions on the rights of the child is encouraging, the government of Uzbekistan has yet to demonstrate a real and sustained determination to eradicate forced child labor which looks to continue unless the political structure of the country is reformed. Until it is, these conventions will share the fate of other unfulfilled ILO conventions ratified by the Uzbek government. The legal framework needed to instantly stop the practices of forced child labor is not missing, rather such laws and conventions are used to obscure the truth of the Uzbek cotton sector. A valid sign of the government's unadulterated intention to eradicate forced child labor would be the preliminary step of putting into operation reforms in the cotton industry.

## CHANGE IN THE UZBEK PEOPLE'S ATTITUDES ABOUT CHILD LABOR

When parents were asked their views on this question under conditions of relative anonymity, they universally state that they would rather keep children in school than have them work in the fields. Pupils unequivocally state they would prefer to be continuing their studies. Some of the parents revealed that compared to other forms of child exploitation in the world, such as sex trafficking, etc., this is not that bad. After all, they worked on a farm (in a shop, in the family's business) when they were children. First, proportional wretchedness underestimates the pain of any individual child or assemblage of children relative to others. Second, enforced child labor in Uzbekistan's cotton fields meets the ILO's classification of the worst forms of child labor, which Uzbekistan itself has agreed to eradicate. Labor in the cotton fields causes injury to health and even death. Children are exposed to these recurrent dangers, and when injured or killed their families are left uncompensated (and even subjected to additional hounding if they pursue any recompense). Third, the work entails harsh educational costs. Children of the countryside miss at least three to four months out of a lyceum/school year in which they already carry a huge disadvantage vis-à-vis their urban peers. If they or their families repudiate involvement, they risk ejection from lyceum/school and will experience at the very least persecution from teachers and school administration. Lastly, Uzbekistan's children do not work for their individual good or that of their families and society, but for the benefit of state and quasi-state structures. Under Uzbekistan's tyrannical totalitarianism, how the earnings from the billion-dollar cotton export industry are consumed remains a secret. Several analysts have seen that those profits do not in reality reach the state budget. They surely fail to do good to the impecunious communities of the countryside.

## ENVIRONMENTAL CONCERNS

Uzbekistan's state-administered cotton industry has also taken its toll on the environment and atmosphere. In the midst of the profound irrigation demands for the hard cash crop; the Aral Sea, once a climate-modifying element in the region, has been reduced to 15 percent of its original size. As a consequence, salinity has increased, killing dozens of species of indigenous fish and devastating Uzbekistan's profitable fishing industry. The cotton fields themselves have been over-irrigated and now suffer from the soaring intensity of soil salinity and corrosion. Cotton monoculture has left Uzbekistan's previously flourishing lands ever more sterile, at times to the point of desertification. The heavy use of insecticides has compounded the ecological crisis, leading to augmented rates of birth defects.

Being the second-largest exporter of cotton in the world; the autocratic state of Uzbekistan maintains a monopoly on the export of cotton. With a scarcely compensated, seasonally caged labor force, a great deal of the income from the cotton is not used in future development projects or for public well-being programs but instead to prop up a parasitic elite that constitutes a dictatorial regime. The cotton that is not exported is used in Uzbekistan's small home textile industry, made up of combined business enterprises sandwiched between the Uzbek state and overseas shareholders. The government of Uzbekistan sells the cotton at 85 percent of the international market price and 43 percent of it is funneled to Asian textile mills. Under similar conditions of forced and underpaid labor, the cotton is transformed into manufactured goods that are then exported to Western countries where it enters the consumer markets. Western consumers are then able to purchase the goods at exceedingly low prices, something that would not be achievable without the exploitation of the workforce in underprivileged countries like Uzbekistan.

# SEVEN

# Critique, Conclusions, and Policy Implications

The cotton industry in Uzbekistan is a contributor to political repression, economic stagnation, widespread poverty, and environmental degradation. Without structural reform in the industry, it will be extremely difficult to improve economic development, tackle poverty and social deprivation, and promote political liberalization in the region. If states, Western governments, and international financial institutions (IFIs) do not do more to encourage a new approach to cotton, the pool of disaffected young men susceptible to extremist ideology will grow with potentially grave consequences for regional stability. Reforming the cotton sector is not easy. Structural change could encourage the growth of an industry that benefits rural farmers and the state equally, but economic and political elites have resisted. Land reform has been blocked in Uzbekistan and Turkmenistan and has moved too slowly in Tajikistan. Farmers still have no permanent ownership of the lands they work and no real say in the choice of crops they wish to grow or to whom they sell their produce. Most researchers who have dealt with the economic transitional countries of the former Soviet Union consider poverty to be a new phenomenon in former socialist states. They perceive poverty as stemming from the disruption of links that bound together the members of the former USSR. While this theory may be correct for some countries, it fails to explain poverty not only in Uzbekistan but also in other Central Asian republics. In reality socialism failed to create a society in which people in the Union enjoyed equal incomes and standards of living nor it did sufficiently achieve income equality during the course of the whole Soviet period. A noticeable differentiation of income also existed between families with European origin and of indigenous people. The ineffectiveness of socialist economics fostered "'equality in poverty'." Although one could argue that the majority of the population belonged to the middle strata, according to their income level, many people lived in poverty while others enjoyed opulent prosperity. Under Soviet rule, ideological dogma impeded the recognition of poverty in the USSR. In fact, in Uzbekistan, about 45 percent of the Uzbek population fell into the category of receiving an income less than the minimum wage. These primarily consisted of rural inhabitants and members of groups such as pensioners, orphans, widows, lone parents (usually female headed households), and disabled people. Over the course of *perestroika* high inflation contributed to an increase in the number of relatively poor people. By 1991, about 75 percent of the total population was living in relative poverty. In the transition period, the problem of poverty became more of a burden since other groups were suddenly confronted with low incomes. Some of these groups once had medium and high levels of income, including doctors, teachers, scientists, government officials, and employees of enterprises, and are now facing financial difficulties associated with structural adjustment.

The case of child labor in Uzbekistan presents distinctive features that set it apart from global patterns. If, in the rest of the world and especially in the South Asia and Africa, the main cause of child labor is poverty, compelling households to send their children out to work, in Uzbekistan the prime movers of this practice are state agents and the particular mode of organization of cotton farming. The use of children ensures maximum profits to the ruling elite, which benefits from the supply of cotton to Western consumers. The use of child labor violates international laws and conventions to which the government of Uzbekistan is a signatory. Children's normal education is interrupted to serve the interest of the small elite who benefit grossly from the high profits from trading cotton on the world market. As a result of forced child labor, children cannot learn in schools and colleges during the academic period, and lag behind in the school curriculum, while some children fall sick from hard work and exhaustion. The Uzbek government continues to use a state-order system of cotton production underpinned by forced labor of children and adults, despite international condemnation including by the United Nations. In the last harvest the government of Uzbekistan intensified the use of forced labor. All government employees—teachers, doctors, and nurses among others—were required to contribute to the cotton harvest under threat of losing their jobs, pay, or benefits. Students and children aged eleven to eighteen made up for the shortfall in younger children as authorities reacted to pressure to end the systematic closure of primary schools evident in previous harvests.

Findings presented underscore the fact that children face numerous challenges while making the transition to adulthood. These challenges call for program interventions at the children, youth, family, and service delivery levels. Key program recommendations emerging from this study are highlighted below.

## EUROPEAN UNION

European Parliamentarians should block all textile agreements with Uzbekistan due to systematic and continuous human rights violations in its cotton sector. Clothes made from the fruits of forced labor of Uzbek children must be boycotted in the whole of Europe. The European Parliament's trade committee in light of the Uzbek government's lack of progress in bringing this practice to a meaningful end should call upon the European Commission to launch an investigation into Uzbekistan's trade preferences with Europe. The bulk of Uzbekistan's cotton, which is picked using forced labor, is exported and ends up in European shops. The majority of Belgium's cotton imports are from Uzbekistan; Uzbekistan is the second-largest supplier of cotton to Italy and Germany; and companies in Turkey, a hub of textile manufacturing for European brands, process Uzbek cotton to a wide extent. Even at a time of crisis, the European Union's five-hundred-million-strong consumer bloc and its position as a leading trade partner is important to a relatively small country like Uzbekistan. The Germans should call for a boycott of Uzbek cotton, while the French National Contact Point for OECD Guidelines for Multinational Enterprises should state unequivocally that the trade in cotton produced by forced child labor constitutes a flagrant violation of the guidelines. In Brussels, the Parliament should also act as the sole body representing Europe's conscience on the issue, as it has also voted overwhelmingly in December 2011 to block the textile agreement between the European Union and Uzbekistan on human rights grounds. In so doing, the Parliament answered the call of European consumers and retailers who, together with human rights NGOs, European schoolchildren, employers, and trade unions and Uzbek Diaspora groups, are increasingly standing up against this practice. But by contrast Europe's trade commission remains silent. Despite Uzbekistan's commitments to relevant ILO conventions banning the use of forced labor and the worst forms of child labor, it has steadfastly refused to implement the ILO's recommendation to invite a high-level tripartite mission to observe the harvest, leaving the ILO's hands tied. Although seemingly remote from Uzbekistan's cotton fields, decisions made in the European Parliament's committee rooms and the corridors of the commission's Berlaymont building do matter.

It is time that the commission investigates the evidence that its rules on trade preferences are being violated by Uzbekistan and remove those benefits. Doing so would not only put into practice the commitment to "make trade work for human rights" in the European Union's Global Human Rights Strategic Framework. It would also ensure trade preference rules are applied fairly, thereby boosting the integrity of this incentive-based system. The mass mobilization of children and adults under threat of penalty—including dismissal from work, the loss of salary, pensions, and welfare benefits—is a serious and systematic violation of core ILO standards and therefore technically meets the European Union's criteria for the suspension of preferences. Yet forced labor on the scale practiced in Uzbekistan is a political, not a technical issue. The European Union needs to use its leverage as a consumer and major trading partner to send the message that systematic use of state-sponsored forced labor is unacceptable. Parliament should ask commissioners to answer the call of citizens from across Europe and open an investigation into Uzbekistan's status as a beneficiary of trade preferences. When political change inevitably comes to Uzbekistan, the Uzbek people will remember if Europe did everything it could to help end their servitude. The answer may well shape their attitudes toward Europe long after President Islam Karimov has left the scene. Until that time, the European Union can at least ensure that its action matches its stated policies on human rights.

## ENACTMENT OF NATIONAL LEGAL PROVISIONS

Although the Government of Uzbekistan has adopted the necessary legal framework for the eradication of forced child labor, both as signatory to international treaties and through domestic legislation, it appears to be in breach of both. The involvement of state parties in the mobilization of child labor for cotton harvests cannot be glossed over. The extended suspension of schooling and the organized large-scale deployment of children on cotton fields can hardly take place without the tacit endorsement and support of the central government and the active involvement of local administrations. Without verifiable benchmarks and systematic monitoring, the NAP may remain a dead letter given the lack of concrete steps to reform the agrarian sector and to lift the current constraints on the operations of the cotton sector. Uzbekistan, with its massive unemployment, was and is entirely able to eliminate forced child labor. The country already has several laws on the books that would, if respected, ban children's forced labor on the cotton harvest. The practice clearly violates Uzbekistan's own long-standing statutes on the rights of children, and the labor code including its provisions on the minimum age and conditions of children's work. Yet despite these laws, over the past decade, the government has only intensified its reliance on forced child labor to bring in the cotton harvest, due to a number of economic and political factors. Among these are the persistence of elements of a command economy in the sector, and constraints on a free labor market. Unemployment touched new heights in Uzbekistan. It is partly due to sharp increase in the working-age population, as is evident by the fact that the labor force increased by 2.3 percent per annum for the years 1985–1991. In 1990 alone more than half a million people were seeking employment, according to the Ministry of Labor. The problem of unemployment is especially acute for the rural population, where the possibilities of creating new jobs have essentially disappeared. All this led to "hidden unemployment"—that is, a situation where enterprises employed more people than actually needed, which in turn led to the declining living standards of the masses. It has also created a big division between rural and urban communities in Uzbekistan. Meaningful steps toward ending the problem are well within the reach of the government of Uzbekistan; the principal obstacle to taking such steps is political will. Under its commitments to the lately signed ILO Convention No. 182, the Uzbek government is obligated to provide the ILO a list of sectors where the worst forms of child labor are found. Public acknowledgment of the problem through public identification of cotton as a target sector is one important precursor to further action.The government of Uzbekistan needs to put in place a comprehensive national action plan to end forced labor in the cotton industry and commit all resources necessary to the implementation of this action plan. Independent journalists and human rights defenders should be having unrestricted access to document the situation during the cotton harvest and addressing and evaluating the situation of forced child labor through annual reports on human rights, human trafficking and labor issues. There should be encouragement to the World Bank and Asian Development bank to resume their efforts to promote reform in Uzbekistan's agricultural and cotton sectors as forced child labor is symptomatic of incomplete reform of these sectors of the economy and the lack of farmers' rights to manage their own assets and products. Multilateral organizations including the International Labor Organization should be encouraged to make public statements through annual and thematic reports about violations of labor and human rights in Uzbekistan's cotton industry.

## LAND REFORMS

The root causes of reliance on coerced labor should be recognized and remedied through a package of reforms that address the vicious cycle of a partially modified command economy, the plight of labor, declining productivity and low incomes in the cotton farming sector. These are factors that fuel rural poverty and erode trust in governance and administration. In order to assess the real magnitude of the employment of minors, an in-depth study is necessary, particularly in regions with high unemployment rates. The major factors contributing to the supply and demand of unemployment of minors need to be better researched as well.The questions of both theoretical and practical importance, which importantly include the effects economic reform efforts had on political liberalization in the countries of Central Asia are to be addressed. While the concepts of structural reform and political change have played key roles, most structural reforms involve changing agencies, changing their powers and authorities, changing their practices and tendencies, and making sure that they do what they are intended to do and not other things. Although large-scale mobilization of labor for cotton harvests, including school-age children, was prevalent during the Soviet period, it would be mistaken to interpret current practices as a mere continuation of the patterns set by collective agriculture. The partial nature of agrarian reforms since independence in 1991, expanding private access to land, on the one hand, without releasing private producers from obligations to meet crop quotas at administratively set prices, on the other, has contributed to the surge of labor outmigration to Russia and Kazakhstan, aggravated problems of labor recruitment and supply, and increased reliance on coercive methods of labor control. Recourse to child labor is symptomatic of the systemic failure of current agricultural policies and the necessity for thorough reform.Thorough reform of the agricultural economy is necessary in order to replace the cheap and easily coerced farm labor that Uzbekistan's schoolchildren now provide. Artificial suppression of purchase prices for agricultural commodities such as cotton will have to be removed so that farmers are able to cover the real market cost of the labor required to grow and harvest such crops. Opening the market in agriculture, and abolishing obligatory state quotas for cotton and wheat may be required to do so.

Throughout, the government has maintained control over large parts of the economy. In agriculture, the authorities control the production and marketing of the two most important crops, cotton and wheat. In industry, extensive support has been provided to keep state-owned enterprises afloat, including through budgetary on-lending, low energy prices, directed credits at favorable terms, and priority access to foreign exchange at the favorable official exchange rate. Regulation is extensive, especially for small- and medium-sized enterprises, which also carry a heavy tax burden. Competition has remained limited in many sectors owing to the dominance of large state-owned enterprises. "Antimonopoly" policies have mainly taken the form of extensive price controls. Various restrictions on business and individuals have been maintained in the financial sector (e.g., on cash withdrawals and on the number of bank accounts).Agriculture has gone through considerable changes in Uzbekistan, and with the completion of the conversion of the *shirkhat* into private farms there is a need to consider the impact on child labor. Though the exercise of structured child labor via schools in conjecture might begin to grow fainter, it may well shift the problem to the family, where children will continue to be deployed on family farms—the least synchronized and potentially most exploitative appearance of child labor.

The limited window of opportunity for bringing in the harvest means that the mobilization of the whole community to bring in the harvest is likely to continue. During the 1990s labor productivity was low, and employment creation in rural areas was hampered by the effectively high taxation of the agricultural sector through the quota system, whereby the prices for cotton and grain were fixed below world price levels. This discouraged investment into technology and more capital-intensive agriculture, which while maybe reducing the number of jobs available would have increased the proportion requiring the kind of skills and experience that children would not have. It has been observed all over the world that land reforms does reduce the socioeconomic inequality present in any society and brings considerable change in agrarian relations. Land reforms serve the twin purpose of reducing the income gap between haves and have-nots and at the same time creating a multiplier effect on the economy of society. Agriculture is considered the backbone of the economy of any country and Uzbekistan was not an exception to it. The decline in agricultural productivity not only has an impact on the rural economy it also affects the working population of the urban areas as well. Uzbekistan does not have big industries but there were many small industrial units that used to be the directly dependent on the agriculture sector, which used to be under the control and management of the central command system during Soviet times. But, when we assess the agrarian reforms that have taken place in Uzbekistan, the situation has not changed a great deal for the rural masses. There are various factors involved that have resulted in the slow pace of agrarian reform. Cotton continues to be the white gold for the government as it earns them major foreign exchange and the government is in no mood to privatize the land. The old pattern of collectivization, which gave workers the security of bread and butter, is no longer existent and the new reforms have not delivered as experts hoped immediately after independence. The main task of the distribution of land among the workers, formation of a class of real owners of capital is not visible still. As a result, workers who used to work during Soviet times on collective farms, cooperatives; state and other collective organizations are still not interested in increasing economic efficiency and production volumes. The agrarian reforms that are taking place are going at a snail's pace. These reforms are mostly determined by the state and don't always mean that they are necessarily beneficial to the rural masses in general and poor in particular. The process of de-collectivization, which started immediately after independence in 1991, has to go a long way and the hurdles in this process can't be over looked.

## RURAL-URBAN DIVIDE

Mass involvement of rural children for the cultivation of cotton during harvest season as well as during other seasons of the year is widening the rural-urban divide, since the schoolchildren put into practice are losing out compared to urban counterparts. Though the autocratic structure of the state prevents any kind of awareness of these inhumane measures violating international commitments, the urban sectors, being better connected to the media and having access to means of communication, could begin to see and expose the state propaganda as lies. The children's involvement in compulsory agricultural labor results in significant losses in schooling and the widening of the rural-urban gap (since the burden falls disproportionately on rural children), a deterioration of human capital with consequences for the intergenerational transmission of poverty, and the consequent erosion of social trust in the state's capacity to provide welfare for its citizens.

This entire scenario has not helped in any way the small peasants or casual laborers who used to be the active participants in the whole agriculture process during the time when collectivization was in operation. These casual laborers mostly find themselves in miserable conditions and have little options available to them to engage themselves in a meaningful economic activity. This section of the rural population is an impoverished lot. It is because of these poor economic conditions that this section of the population is migrating from rural Uzbekistan to Russia and other neighboring countries for want of a livelihood. These laborers generally work as daily wage earners in the construction field, particularly in Russia, which of course is seasonal. During the Soviet rule skilled workers, managers, and planners were usually from Moscow; now it is blue collar-workers from Uzbekistan who migrate to Moscow. This has quite a drastic effect on the overall rural social structure of Uzbekistan, which is far more sedentary than the other four republics. In Soviet times working in *Kholkhoz* was mandatory, and to leave the village one needed a genuine reason. Even after having good reason to leave a village the authorities seldom allowed people to move from one village to another even for a very short period of time. That is probably one of the reasons that the traditional joint family remained a dominating feature of this society. But, things are other way round now; people do not want to move but they are compelled to do so purely because of economic reasons.

Those who benefit from Uzbek forced child labor, the Uzbek comprador elite and the First World, are a global minority. In contrast, the Uzbek masses are part of a larger majority, the vast Third World masses. Hundreds of millions of children around the world are compelled to work, presumably in commodity exchange industries. Adding to this are the world's exploited adults, those languishing in vast urban slums and subsistence communities under constant threat of being kicked off the land.

Together, the vast Third World masses pose a serious threat to the system: they carry great potential and a historical responsibility.

## CROP MANAGEMENT AND CROP ROTATION

The landless laborer's daily or monthly wages are determined by the contractor. The sense of belongingness is missing, as the farmer can neither decide which crop to grow nor sell the produce in the open market. This alienates peasants from the land, which has all the negative effects on the overall production process. Because this process is inherently incompatible with the basic desire of human nature—that is, the sense of owning things. Human beings have an innate desire of owning stuff. Once this feeling of owning the very land on which they work is missing, everything is affected, whether it is operating these fields efficiently, managing irrigation, or simply putting their best efforts in order to have some additional income. So, all these things make it really difficult for any researcher to say whether the process of de-collectivization has at all taken place or it has only changed its form. Yes there are certain instances where the poor peasants enter into a contract with the district administration, what they call a "Sharikat." But, these poor peasants are again dependent upon the big farmers when it comes to hiring of the agricultural technology, which is required at different stages of operation, whether it is at the time of ploughing the fields or at harvesting time.

## ADVANCED COTTON VARIETIES AND AGRICULTURAL METHODS

The advanced methods in the operation of agriculture in the contemporary world are very significant in increasing production as well as freeing the manual labor force for other more lucrative jobs in the job market. And equally important is the introduction of advanced varieties of crops for producing quality and quantity crops. Cotton, which is cultivated by machines throughout the developed world, is cultivated by hand and often with the use of innocent schoolchildren in Uzbekistan. The same could well be put into practice in Uzbekistan as was practised by Russians. Advanced varieties of cotton are available now on the market and could be used for increasing the yield as well as the quality of the crop, as in the developed countries. One of the most important problems of agricultural development in Uzbekistan is rural overpopulation and high population pressures on land. The government should focus on the problem of increasing the occupational level of the rural population. Uzbek agriculture faces several severe problems related to the efficiency of labor force usage, including growing indicators of labor requirements, decrease of the level of mechanization, inefficient use of labor, and so on. To increase the efficiency of labor in agriculture it is necessary "to consider the issues of improving the efficiency of agricultural production through the elimination of the cost approach to the use of the labor force, through the decrease of manual and unskilled labor on the basis of the improvement of technical equipment of agricultural production, extricating the superfluous people, and the redistribution of that labor into the processing industry and into the sphere of agricultural and community service. The main stress should be on the purposeful management of extrication and redistribution as a way to form the optimal occupational structure of the population.

In Uzbekistan agriculture being the largest sector of the economy. As measured in terms of GDP, employment and export potential makes it indispensable for the economy as a whole and more so for the rural development. This goes without saying that it is the leading sector and forms the basis of the entire national economy. Specifically it plays a vital role in the economy by engaging 60 percent of the rural population, which is the basis for the development of leading industries and being one of the main sources of export resources and hard currency earnings. The agricultural sector accounts for 12.2 percent of production assets, 34.4 percent of employment, 70 percent of trade and 30.1 percent of GDP, and 60 percent of exports. The country produces 3.5–3.7 million tons of cotton every year of which 1–1.2 million tons of cotton fiber is of world standards and is considered a world leader in developing high-yield strains of cotton. The cultivation of cotton makes Uzbekistan one of the world's biggest regions in terms of the use of irrigation farming. Despite using more than 50 percent of the river basins of the Amu Darya and Syr Darya (rivers), which accounts for only 20percent of the water needed for irrigation, and the other 80 percent of water for irrigation comes from Kyrgyzstan and Tajikistan; Uzbekistan has the advantage of a warm climate and a long growing season. In the Soviet period, these conditions offered high and reliable yields of crops with specialized requirements. Soviet agricultural policy applied Uzbekistan's favorable conditions mainly to cotton cultivation. As Uzbekistan became a net exporter of cotton and a narrow range of other agricultural products, however, it required large-scale imports of grain and other foods that were not grown in sufficient quantities in domestic fields. Hence the agrarian system of the whole of Central Asia had been shaped in a particular way such that the present political rulers are trying to make the best use of the collectivization that was the hallmark of the Soviet legacy and at the same time taking note of the transitional phase of their economy they find themselves in. However, the main economic reality of the regional life remained for a very long time a poorly developed agriculture. Many structural distortions were introduced in the economy of the region, which was favorable for each new colonial capitalization. Therefore the contribution of agriculture remained important even when Central Asia entered a new phase of industrial development in the 1960s. Since then the emphasis shifted to build a more complex industrial structure. This was also a period of qualitative development of industry, based on the scientific and technological revolution that accelerated development of those branches of industry that held out the prospects of a qualitative transformation. Uzbekistan, for example, developed such industrial branches as mechanical engineering, iron and steel, non-ferrous metal industry, air craft building, chemical industry, gas extraction, fuel and energy industry, and gold mining, and accordingly exported cotton, textile machinery and other complex equipment. Despite all these developments in agriculture, we must not fail to underline that most of Central Asia is arid and a considerable part is covered with desert, which leaves very few areas of the region suited for agriculture.

Transforming the cotton industry in Uzbekistan will not be simple. Structural reforms are needed to stimulate the growth of the cotton productivity, being beneficial for the cotton industry, farmers, and ordinary people alike. Cotton monoculture is more disastrous for the future of Uzbekistan than the tons of heroine that are regularly transported throughout the Central-Asian region. Even though the international community has invested millions of dollars in antinarcotics programs, it has done far too little to fight against the negative effects of the Uzbek cotton industry. Achieving real changes in the cotton sector will take time, but true reform of this sector could provide more stability in this strategic country than almost anything else the international community could offer.

## POVERTY ELEVATION PROGRAMS

Child labor is not simply a response to rural poverty at the household or community level but is an intrinsic feature of the current operations of the cotton sector in Uzbekistan and part and parcel of its methods of labor control. As a consequence, the process of eradication of child labor can only become feasible as a component of a broader package of reforms in agricultural policies. Currently, hundreds of thousands of children appear to be involved in harvesting activities and are responsible for a substantial proportion of the cotton harvested. A path of agrarian reform that releases primary producers from the administrative dictates of central government, that stimulates the growth of agricultural incomes, and that diversifies the economy in ways that promote alternative employment would obviate the need for coercive means of labor control, including recourse to child labor, and restore citizens' confidence in their government's ability to safeguard their welfare.

## TRADE ORGANIZATIONS

Reducing the number of controlling actors in the cotton chain will free the share of export revenues currently absorbed by these persons. This will help bridge the gap between the procurement price of cotton and its export price. Central Asian cotton is traded internationally by major European and U.S. corporations; its production is financed by Western banks, and the final product ends up in well-known clothes outlets in Western countries. But neither the international cotton trading companies nor the clothing manufacturers pay much attention to the conditions in which the cotton is produced. Nor have international organizations or international financial institutions done much to address the abuses. U.S. and EU subsidy regimes for their own farmers make long-term change more difficult by depressing world prices. Since gaining independence in 1991, Uzbekistan's authoritarian government has become more reliant on the use of forced child labor to harvest cotton due to lack of investment or economic reforms in agriculture. The international community paid little attention to this issue amid Uzbekistan's other severe human rights problems until local activists in 2004 and 2005 began to call on the world to boycott the cotton, harvested by children, which is exported and sold around the world. These calls began to have an effect in 2007 and 2008, with international brands and retailers including Tesco, Walmart, Target, Levi Strauss, Gap, Limited Brands, and Marks and Spencer agreeing to ban Uzbek cotton from their supply chains. In 2008 and early 2009, Uzbekistan signed two ILO conventions against child labor in response to this international campaign. Uzbekistan has sought to counter the economic pressure applied on Tashkent through the boycott of Uzbek cotton by leading U.S. and Europeans clothing companies by reorienting its cotton exports to Asian and Middle Eastern markets. Evidence that the government is feeling the effects of the boycott is inconclusive so far. At the Fourth Annual Cotton Fair, held in Tashkent, Uzbek officials signed deals worth approximately $1 billion to export nine hundred fifty thousand tons of cotton fiber. The chief purchasers included China, India, Pakistan, Bangladesh, South Korea, and the United Arab Emirates, according to the pro-government website Gazeta.uz.

## STATE/POLITICAL STRUCTURE

The state of affairs that Uzbekistan finds itself in at present is a direct consequence of long colonial rule. The current ruling political elite are again the same people who used to be at the helm of affairs during Soviet times. So their policies cannot be much different from what they used to be in the past. Experts believe that it will really be very difficult for the economy to progress as long as bottlenecks to the process of democracy are not removed. Once people in general and civil society in particular, which are the actual stake holders, are cut out of the decision-making process, things on the ground are unlikely to change. Some scholars believed that once these republics were liberated from Soviet rule there would be a resurrection of pre-Soviet identities in the form of tribal divisions and Islamic fundamentalism, or the violent rejection of the Soviet legacy in the form of nationalism and ethnic conflict. But after fifteen years of independent rule, with the exception of Kyrgyzstan, Central Asia's transition from Soviet rule has been by and large peaceful. It goes without saying that inter-ethnic unrest in the Osh region of Tajikistan in 1990 and civil riots in Dushanbe in 1992 forced Karimov to change the policy toward Islam. Although the Kirghiz constitute just under 1 percent of Uzbek population according to the 1989 census, the majority of whom live in the Kirghiz part of the Fergana valley, these developments in the immediate neighborhood led to the closing of borders and a general clamp down on Uzbek opposition. How well this approach of government has succeeded is debatable, but the fact of the matter is that we have not witnessed any major incidences of warfare between ancient clans or tribes. We also have not witnessed the rise of Islamic fundamentalism in Uzbekistan in comparison to other parts of the former Soviet Union, where there have been large-scale nationalist uprisings. The reassertion of national identity in Uzbekistan affirms that the major social and political events are dressed up in a costume of Uzbek national identity, which is always directed to promote a particular vision. The underlying presumption is that a strong concept of national culture will ensure the loyalty of the people to the regime. However, it remains to be seen how long people will remain attached to this kind of spoon feeding from the different organs of the state. As everyone is aware, political communication in Uzbekistan consists mainly of the one-way transmission of ideology from state to citizen. Cultural elites are lured and in some cases coerced to fall in line, particularly the media, which are state controlled. In this kind of scenario it becomes very difficult for any political party to show dissent toward any of the government policies, whether they are economic or political in nature. It is precisely because of these circumstances that very few voices are raised against any policy issue related with socioeconomic development particularly agrarian reforms.

In Uzbekistan gangs forcibly remove hundreds of thousands of children from schools, ordering them to pick cotton in the searing heat and live in squalid conditions on pitiful wages. Blended by manufacturers thousands of miles away, Uzbek state-controlled cotton is sold to the world's biggest retailers, making the repressive regime the third-largest exporter of "white gold" and earning the government $1billion. But, in what has been described as a major breakthrough, a decision by some of the world's biggest clothing businesses has forced the Uzbek government to sign International Labor Organization conventions that commit the country to stop using child labor in its state-sponsored industry. This is a major step forward. Virtually nothing persuaded the government to change course. But the actions of retailers and campaigners are definitely now having an impact. The key question that remains is whether the Uzbek government will implement the conventions. They need to allow independent monitoring and work with civil society, which are basic requirements of the conventions they have signed and ratified. They are as yet not doing this, so the jury is out. The Uzbek government disputes that its cotton industry sponsors forced child labor, saying that this claim has been spread by cotton-producing countries that are losing market share. It adds that its monitors ensure the best international standards are adhered to. The London insurance market plays a key role in providing cover for the repressive military junta's port and aviation interests. Without insurance, campaigners say, the junta would be starved of cash and so be more likely to embrace the international community.

There is no denying the fact that all the Central Asian republics were used as an appendage by the Russia in the past. These Republics were exploited, whether it was for their natural resources or agriculture, by converting them to monoculture regions. It was the central command system that changed most of the agricultural land into cotton fields so that their industry could be supplied with cheaper raw material on a regular basis and later on send the finished products back for marketing. They were not in any sense different from other imperialists; they came, conquered, and exploited all these republics by turning them into colonies. But there is another aspect to it, as has been the case with most of the imperialists—that is, to help these republics to grow from an industrial point of view. This is also a hard fact. Before 1917 this region had all the characteristics of a backward society with low levels of literacy and no traces of industrialization. But by the time of the Soviet collapse, the region's infrastructure could match that of any developing country in the world, whether in means of transportation/communication or developing human resources. This is evidenced by the fact that Uzbekistan does have 98 percent literacy at present. The Soviets' great emphasis on education helped this region in general and Uzbekistan in particular to improve upon its human resource potential. This is the basis for the future development of this region. There is no doubt that at present these economies are in shambles and the rural poverty is obvious, but the people, particularly the youth, are now well equipped to migrate to any part of the world, which they often do, particularly to Europe. However, if the government wants to stop emigration from the countryside, it must really push forward agrarian reforms that are long overdue. The government has not been able to utilize the agriculture resources because of the slow pace of these reforms. Nobody can deny the fact that the government earns huge amounts of foreign exchange from exporting cotton to Russia and other European countries. So it has a genuine fear of losing out on this foreign exchange if land privatization takes place. However, there is an urgent need to maintain and even increase diversity; not just biological diversity, but also diverse lifestyle, for which this region has been famous since time immemorial. The homogenization of the landscape has to be avoided because there are tremendous repercussions in the degradation of the Aral Sea on the one side and salinization of land on the other. One has to look for other options for sustainable development that can maintain the diversity of crop cultivation and lifestyle as well. The answer is to integrate these diverse lifestyles into sustainable development programs for the common benefit of nature and human society. Achieving sustainable development will require changes in consumption patterns in many instances; this will necessitate reorientation of the production and consumption patterns that have developed in industrial societies and have been emulated throughout much of Central Asia in general and Uzbekistan in particular.

## INTERNATIONAL LABOR ORGANIZATION (ILO)

The government of Uzbekistan should publicly renounce forced child labor in the cotton industry of Uzbekistan and put in place a comprehensive national action plan to end forced labor in the cotton industry; commit all resources necessary to the implementation of this action plan; invite an ILO high-level mission to Uzbekistan during the harvest season; implement its commitments under ILO Conventions No. 29, No. 105, and No. 182; and allow independent journalists and human rights defenders unrestricted access to document the situation during the cotton harvest. The field interviews clearly show that parents, children, teachers, and even farmers whose livelihoods are aided all deplore the forced labor of children. This suggests the root causes of the problem in Uzbekistan are not poverty, tradition, or lack of schooling, as sometimes is the case in other countries. The problem is rooted in the nature of the industry and state control over rural populations. Respondents noted that the large numbers of unemployed people in their communities could be mobilized to pick cotton if paid a reasonable wage; thousands of Uzbek citizens migrate to neighboring countries each season to do the very same work that they disdain at home because it is so poorly paid. Children, however, are more easily subject to state pressure. Efforts that suggest the need to educate farmers, parents, or teachers are misguided and risk deflecting attention from the problem's real root causes. The international community needs to vastly increase its efforts to monitor the cotton harvest and hold the government of Uzbekistan accountable to end forced adult and child labor.

There is a pressing need for adequate data and methodologically sound instruments to monitor patterns of child labor in the cotton sector of Uzbekistan. Specialist UN agencies such as UNICEF and the ILO must, in dialogue with the government of Uzbekistan, break with the precedent set by flawed surveys by various agencies and redouble their efforts to achieve a reliable data base as a guide to future policy. There is no lack of regulation concerning child labor, and as the Uzbekistan government's 2005 report to the CRC Committee points out, the "labor legislation norms are more favourable than the minimum norms envisaged by the international legal acts on labor and by the ILO Conventions." The problem with regulation is the lack of mechanisms for implementing the labor legislation; labor inspectors need to be better focused on the violations taking place. The main issue for regulation, however, is to better focus on the issue of child labor in bringing in the cotton harvest. Currently there is no regulation addressing this in education or labor laws. This leaves local authorities to be the arbiter of the duration of children's cotton harvesting work, its conditions, and payment. The importance to the economy of this practice must be more openly debated, and a reasoned discussion is called for to see how children might become involved in ways that are not detrimental to their health and education. The incentive structure built into the quota system must be examined so that local authorities are penalized rather than being rewarded for exploitative forms of child labor.

## DEVELOPED NATIONS

The Republic of Uzbekistan is among the eligible beneficiaries for the US Generalized System of Preferences (GSP) under the Trade Act of 1974, which also lists cotton and cotton-made products as commodities and goods subject to GSP rules. Uzbekistan is listed as one of the countries to be included in the Department of Labor's annual report on trade beneficiary countries' implementation of international commitments to end the worst forms of child labor. There are significant and growing concerns regarding Uzbekistan's deteriorating human rights record, both directly and indirectly linked to cotton production. The government of Uzbekistan has not only failed to enforce its laws against forced and compulsory labor, but also continues to deny the existence of the problem. When asked to comment on forced child labor in the cotton industry the Uzbek authorities often call it a patriotic act by the Uzbek youth to collectively gather one of the main export productions of country—cotton.

People's aspirations and hopes are largely associated with the future of their children and achievements within their individual families. Regarding the national level, the aspirations and hopes of most people are connected with economic recovery and political stability. This corresponds to a discontent with the present economic situation and government policies. The state is still perceived by people as the most legitimate organization for providing their needs. Fundamental respect toward the state as a legitimate representative institution is still maintained in the minds of the people. Yet, increasingly people call for a faster and larger degree of democratization in society. The aspiration for a more open and democratic system is exemplified by declining public confidence in state institutions providing for public social welfare, a low level of public electoral activity, and a general pessimism regarding ordinary people's ability to influence political decisions. Economic growth in the country is still seen as the most important objective, leading to higher incomes and a more developed political system. However, the public belief that economic growth is the only way toward democratization might be fading. In other words, the public belief that a strong developmental state represented by a strong executive power can lead to high economic growth and then to the democratic reforms may be weakening, as shown by public frustration both with governmental economic policies and with the state of democracy in the country. On the contrary, there is a tendency in responses to support wider public participation in political life, democratic elections, enforcement of individual human rights, and eradication of corruption among governmental officials. Although not directly addressed in the polls, such public support for wider democratization in the country may be seen as an indication, however tentative, of public preference for a more open economic system as a possible path to higher economic growth. Such a shift in public views on development and democratization could be largely explained by the recent experience of economic reforms, which, in more than twenty years of independence, have not produced much in terms of higher personal incomes or employment opportunities. What is obvious, though, is that the reserves of public confidence that government was granted after independence have critically shrunk and people are increasingly impatient with governmental failure in economic reforms.

## INFORMATION BASE

The need for an open debate is essential for reasonable regulation to be developed, and the way to start is greater transparency and more data around the issue. The question put by the CRC Committee concerning "the number of children involved in labor who are under 16" unfortunately went unanswered in the government's reply. A larger information base is required in Uzbekistan, which will assess the correct proportion of child labor in the country. Surveys carried out in seasons other than the harvest season have led to an underestimation of child labor.

This study provides a broad picture of cotton cultivation and child labor in Uzbekistan. At the same time, findings have raised a number of issues that require further investigation, particularly with regard to the determinants and consequences of children's behaviors and practices in the process of cotton cultivation. The study highlights the need for further research in terms of formative research that explores in greater depth factors impeding successful transitions to adulthood, including enrolment in school and school completion, entry into the labor force, initiation of sexual activity, and marriage and parenthood. Research is also needed that explores the role of human rights agencies and the media, socialization practices, young people's access to information and services, and the ways in which these factors contribute to or impede children's ability to make successful transitions. Moreover, a better understanding is required of the factors underlying children's agency and the expression of unequal and equal gender role attitudes. Also required are efforts to refine measures of agency as applicable to children. A general research recommendation is the urgent need for prospective or panel studies that follow a cohort of children, youth, and adolescents at regular intervals up to age thirty-five. Prospective study designs would enable researchers to take a life-course approach, identify with compelling data the factors responsible for healthy transitions to adulthood, and point to the ways in which the situation and experiences of children and youth in adolescence influence their life course at later ages.

Operations research is also needed. Urgently needed, therefore, are carefully designed and rigorously tested intervention models that not only pay attention to the content and delivery of the intervention but also measure effectiveness and acceptability—in short, that will enable a shift from the implementation of *promising* to *best* practices in addressing issues. In order to inform the field, multiple inputs are required. Ultimately, research is needed that monitors the scaling up of successful interventions in terms of their impact on child laborers' and people's lives.

This study has documented, for the first time, the multifaceted situation of child laborers in the cotton industry of Uzbekistan. The study alerts us to the many challenges that Uzbek schoolchildren who are forced to work in the cotton industry must face and their ability to make a successful transition to adulthood. Programs must recognize the needs and requirements of the children and interventions and delivery mechanisms should be appropriately tailored to meet their needs. Evidence presented here provides not only a blueprint for the programming needs of child workers in Uzbekistan but also a baseline by which to measure the impact of programs intended to address their needs.

# Glossary of Terms

*hokim*—governor of province

*khashar*—popular tradition of aiding one's neighbors, relatives, or the local community. In Soviet times authorities began to use khashar to describe forced labor on days free from one's main work. Most often this consisted of street sweeping and other cleaning of public places.

*kolkhoz, pl. kolkhozy*—collective farm (Russian)

*mahalla*—neighborhood

*oblast*—province (Russian, viloyat in Uzbek)

*shartnoma*—contract

*shirkat*—agriculturalcooperativecloselycorrespondingSovietcollective farms (sovkhoz or kolkhoz), but usuallywith less land than before and with a new name

*shtab*—daily meetings

*som*—Uzbekistan's national currency

# Bibliography


Abdullaev, Iskandar, Mark Giordano, and Aziz Rasulov. 2007. "Cotton in Uzbekistan: Water and Welfare." Proceedings of The Cotton Sector in Central Asia: Economic Policy and Development Challenges, conference held at SOAS, University of London, November 3–4, 2005 (coordinated by Deniz Kandiyoti).

ADB (Asian Development Bank). 2009. *Asian Development Outlook 2009 Update: Broadening Openness for a Resilient Asia.* Available at http://www.adb.org/Documents/Books/ADO/2009/Update/default.asp.

Ahmad, S. Akbar, and Hastings Donnan (eds.). 1994. *Islam, Globalization and Postmodernity*. London: Routledge.

Akiner, Shirin. 1994. *Political and Economic Trends in Central Asia.* London: British Academic Press.

Albrow, Martin. 1996. *The Global Age: State and Society beyond Modernity*. Cambridge: Polity Press.

Alexander, Christopher. 2009. *A Carpet Ride to Khiva: Seven Years on the Silk Road.* London: Icon Books.

Allworth, Edward A. 1990. *The Modern Uzbeks: From the Fourteenth Century to the Present: A Cultural History*. Stanford, CA: Hoover Institution Press.

Amelina, A. 1999. "The (Not So) Mysterious Resilience of Russia's Agricultural Collectives." *Transition* 10, no. 6: 19–21.

Anker, R. 1985. Introduction. In V. Bodorova and R. Anker (eds.), *Working Women in Socialist Countries: The Fertility Connection*. Geneva: International Labour Office.

Annan, A. Kofi. 2001. *We the Children*. New York: UNICEF.

Arsel, Murat, and Max Spoor (eds.). 2010. *Water, Environmental Security and Sustainable Rural Development: Conflict and Cooperation in Central Eurasia*. London: Routledge.

Ashirova, Aygul. 2009. *Stalinismus und Stalin-Kult in Zentralasien: Turkmenistan 1924–1953*. Stuttgart: Ibidem.

Baffes J. 2007. "Cotton-Dependent Countries in the Global Context." In D. Kandiyoti (ed.), *The Cotton Sector in Central Asia: Economic Policy and Development Challenges*. London: The School of Oriental and African Studies.

Baley, William. 2009. *The Afghanistan Wars*. 2nd edition. Basingstoke, UK: Palgrave Macmillan.

Balzerm, Marjorie (ed.). 2010. *Religion and Politics in Russia: A Reader*. London: M.E. Sharpe.

Banuazizi, Ali, and Myron Weiner. 1994. *The New Geopolitics of Central Asia and Its Borderlands*. Bloomington: Indiana University Press.

Barkey, Henri J. 1995. *New Countries in an Age of Constraints: The Emergence of Central Asia*. Washington, DC: National Council for Soviet and East European Research.

Beckwith, Christopher. 2009. *Empires of the Silk Road: A History of Central Eurasia from the Bronze Age to the Present*. Princeton, NJ: Princeton University Press.

Beyer J., and Z. Inogamova. 2010. *Baiyz Apanyn Zhashoo Tarzhymaly* [The life history of Baiyz Apa] (in Kyrgyz). Bishkek: Gulchynar.

Bhargava, Pramila H. 2003. *The Elimination of Child Labour: Whose Responsibility ?* New Delhi: Sage Publications.

Blank, Stephen. 1995. *Energy, Economics, and Security in Central Asia: Russia and Its Rivals*. Carlisle Barracks, PA: Strategic Studies Institute.

Boal, A. 1985. *The Theatre of the Oppressed*. London: Pluto Press.

Bovingdon, Gardner. 2010. *The Uyghurs: Strangers in Their Own Land*. New York: Columbia University Press.
Burra, N. 1995. *Born to Work: Child Labour in India*. Oxford: Oxford University Press.
Caldwell, Melissa (ed.). 2009. *Food and Everyday Life in the Post-Socialist World*. Bloomington: Indiana University Press.
Canfield, Robert L., and Gabriele Rasuly-Paleczek (eds.). 2010. *Ethnicity, Authority, and Power in Central Asia: New Games Great and Small*. London: Routledge.
Carley, P. M. 1989. "The Price of the Plan: Perceptions of Cotton and Health in Uzbekistan and Turkmenistan." *Central Asian Survey* 8, no. 4: 1–38.
Carlisle. 1996. "Forced Student Labour in Central Asia's Cotton Fields." *Surviving Together*. 14, no. 4.
Cassandra, Cavanagh. 1992. "Uzbekistan Re-examines the Cotton Affair." *REE/RL Research Report* 1, no. 37 (September).
Castells, Manuel. 1996. *The Rise of the Network Society* (*The Information Age*, Vol. 1). Cambridge, MA: Blackwell.
Catholic New Times. 2005. *Child Labour in Central Asia*.
Central Asian Countries Initiative for Land Management. 2009. *CACILM Multi-Country Program Framework Support Project (CMPF-SP) Year End 2008 Report*, Uzbekistan, Tashkent 2009. Available at http://www.adb.org/Documents/CACILM/NSECUZB-2008.pdf.
Chen, K., S. Chiang, and S. F. Leung. 2003. "Migration, Family, and Risk Diversification." *Journal of Labor Economics* 21, no. 2.
*CIA World Factbook*. Available at https://www.cia.gov/library/publications/theworld-factbook.
Commercio, Michele E. 2010. *Russian Minority Politics in Post-Soviet Latvia and Kyrgyzstan: The Transformative Power of Informal Networks*. Philadelphia: University of Pennsylvania Press.
Committee on the Rights of the Child. 2006. *Forty-Second Session: Consideration of Reports Submitted by State Parties under Article 44 of the Convention. Concluding Observations: Uzbekistan*. Geneva: Committee on the Rights of the Child.
Coudouel, A., and S. Marnie. 1999. "The Mahalla System of Allocating Social Assistance in Uzbekistan." In *Central Asia 2010: Prospects for Human Development*. New York: UNDP.
Craumer, P. 1995. *Rural and Agricultural Development in Uzbekistan*. London: Royal Institute of International Affairs.
Critchlow, James. 1991. *Nationalism in Uzbekistan: A Soviet Republic's Road to Sovereignty*. Boulder, CO: Westview Press.
Cummings, Sally (ed.). 2010. *Symbolism and Power in Central Asia: Politics of the Spectacular*. London: Routledge.
Cunningham, H. 1991. *The Children of the Poor*. Oxford: Blackwell.
Davies, R. W. 1980. *The Socialist Offensive: The Collectivization of Soviet Agriculture 1929–1930*. London: Macmillan.
Deshpande R. 2006. "Land Reform and Farm Restructuring in Central Asia: Progress and Challenges Ahead." In S. C. Babu and S. Djalalov (eds.), *Policy Reforms and Agriculture Development in Central Asia*. New York: Springer.
di Cosmo, Nicola, Allen Frank, and Peter Golden (eds.), 2009. *The Cambridge History of Inner Asia: The Chinggisid Age*. Cambridge: Cambridge University Press.
Dicken, Peter. 1998. *Global Shift: Transforming the World Economy*. 3rd ed. London: Chapman.
Dosybiev, Daur. 2005. "The Dark Side of White Gold in South Kazakhstan." Proceedings of The Cotton Sector in Central Asia: Economic Policy and Development Challenges, conference held at SOAS, University of London, November 3–4, 2005 (coordinated by Deniz Kandiyoti).
Environmental Justice Foundation. 2005. *White Gold: The True Cost of Cotton*. London: Environmental Justice Foundation.

Environmental Justice Foundation. 2009. *Somebody Knows Where Your Cotton Comes From: Unraveling the Supply Chain*. London: Environmental Justice Foundation.

Environmental Justice Foundation. 2010. *Slave Nation: State Sponsored Forced Child Labour in Uzbekistan's Cotton Fields*. London: Environmental Justice Foundation.

ESCOR. 1999. *Rural Domestic Economy and Female Labour Supply in Uzbekistan: Assessing the Feasibility of Gender-Targeted Micro-Credit Schemes*. London: Final report, DFID, Unit, Grant no. R6978.

Eurasianet. 2009. *Tajikistan: Cotton Harvest Relies Heavily on Child Labor*. Eurasia Insight, March 11, 2009. Available at http://www.eurasianet.org/departments/insightb/articles/eav110309.shtml.

Eurasianet. 2009. *Eurasia Insight, Uzbekistan: Arbitrary Land Distribution Feeds Discontent*, posted September 15, 2009. Available at http://www.eurasianet.org/departments/insightb/articles/eav091509.shtml.

Ewans, Martin. 2010. *Securing the Indian Frontier in Central Asia: Confrontation and Negotiation, 1865–95*. London: Routledge.

Ferdinand, Peter. 1994. *The New Central Asia and Its Neighbours*. London: Royal Institute of International Affairs.

Ferghana News. "Uzbekistan: Children Continue to Work on Cotton Fields despite Official Ban." *Ferghana.ru*, September 16, 2009. Available at http://enews.ferghana.ru/article. php?id=2466.

Fierman, W. 1997. "Political Development in Uzbekistan: Democratization." In K. Dawisha and B. Parrott (eds.), *Conflict, Cleavage and Change in Central Asia and the Caucasus*. Cambridge: Cambridge University Press.

Fitzpatrick, S. 1995. *Stalin's Peasants: Resistance and Survival in the Russian Village after Collectivization*. New York: Oxford University Press.

Foltz, Richard, 2010. *Religions of the Silk Road: Premodern Patterns of Globalisation*. 2nd ed. Basingstoke, UK: Palgrave Macmillan.

Freedom House. 2000. *Freedom in the World: The Annual Survey of Political Rights and Civil Liberties, 1999–2000: Uzbekistan Country Report*.

Fyfe, Alec. 2007. *The Worldwide Movement against Child Labour: Progress and Future Directions*. Geneva: International Labour Office.

George, Julie A., 2009. *The Politics of Ethnic Separatism in Russia and Georgia*. Basingstoke, UK: Palgrave Macmillan.

Gerasimov, Ilya, Jan Kusber, and Alexander Semyonov (eds.). 2009. *Empire Speaks Out: Languages of Rationalization and Self-Description in the Russian Empire*. Lieden: Brill.

Gerner, Kristian, and Stefan Hedlund. 1989. *Ideology and Rationality in the Soviet Model: A Legacy for Gorbachev*. London: Routledge.

Gervers, Michael, Uradyn Bulag, and Gillian Long (eds.). 2008. *The Exploitation of the Landscape of Central and Inner Asia: Past, Present and Future*. Toronto: University of Toronto Press.

Gibbons, D. Elizabeth, Friedrich Huebler, and Edilberto Loaiza. 2005. *Child Labour, Education and the Principle of Non-Discrimination*. New York. UNICEF.

Giddens, Anthony. 1990. *The Consequences of Modernity*. London: Polity Press.

Giustozzi, Antonio (ed.), 2009. *Decoding the New Taliban: Insights from the Afghan Field*. New York: Columbia University Press/London: Hurst and Co.

Giustozzi, Antonio, 2009. *Empires of Mud: Wars and Warlords of Afghanistan*. London: Hurst and Co.

Gleason, G. 1991. "The Political Economy of Dependency under Socialism: The Asian Republics in the USSR." *Studies in Comparative Communism* 24, no. 4: 335–53.

Gleason, Gregory. 2008. "The Idea of Greater Central Asia." *Journal of Central Asian Studies* 17, no. 1.

Glenn, John. 1999. *The Soviet Legacy in Central Asia*. New York: Palgrave Macmillan.

Goldin, J., and K. Reinert. 2006. *Globalization for Development: Trade, Finance, Aid, Migration, and Policy*. Washington, DC: The IBRD/The World Bank.

Government of Uzbekistan. 2007. *National Plan of Action for Securing Child Welfare in Uzbekistan*, Tashkent: Government of Uzbekistan.

Government of Uzbekistan. 2007. *Welfare Improvement Strategy Paper: Full Strategy Paper for 2008–2010*, Tashkent: Government of Uzbekistan.

Goyal, P. K. 2005. *Street Children and the Child Labour*. Delhi: Vista International Publishing House.

Grossman, G. 1989. "Informal Personal Incomes and Outlays in the Soviet Urban Population." In A. Portes, M. Castells, and L. Benton (eds.), *The Informal Economy: Studies in Advanced and Less Developed Countries*. Baltimore: Johns Hopkins University Press.

Group of Human Rights Defenders and Journalists of Uzbekistan. 2008. *Forced Child Labor in Uzbekistan's 2007 Cotton Harvest: Survey Results*, Tashkent.

Group of Human Rights Defenders in Uzbekistan. 2009. *Pick All the Cotton: Update on Uzbekistan's Use of Forced Child Labor in 2009 Harvest*, in collaboration with the International Labor Rights Forum, Washington DC: ILRF. Available at http://www.laborrights.org/sites/default/files/publications-and-resources/UzbekCottonFall09Update.pdf.

Group of Uzbek Human Rights Activists and Researchers. 2009. *"We Live Subject to their Orders": A Three-Province Survey of Forced Child Labor in Uzbekistan's 2008 Cotton Harvest*, in partnership with the International Labor Rights Forum, ILRF, Tashkent-New York.

Guadagni, Maurizio, Martin Raiser, Anna Crole-Rees, and Dilshod Khidirov. 2005. *Cotton Taxation in Uzbekistan: Opportunities for Reform*, ECSSD Working Paper No. 41, Washington, DC: World Bank.

Gufronov, D., and R. Mizobekova, "Podmochennyi urozhai." *Azia Plus*, May 7, 2009. Available at http://www.asiaplus.tj/articles/96/3490.html.

Hagemann Frank, Diallo Yacouba, Alex Etienne, and Farhad Mehran. April 2006. *Global Child Labour Trends 2000 to 2004*. Geneva: International Labour Office.

Halimova, Nargis. 2007. "The Cotton Sector in Tajikistan: From Macro-Economic Impact to Social and Environmental Consequences." Proceedings of The Cotton Sector in Central Asia: Economic Policy and Development Challenges, conference held at SOAS, University of London, November 3–4, 2005 (co-ordinated by Deniz Kandiyoti).

Halwai, Mushtaq A. 2008. *Ethnicity in Uzbekistan: A Study in Culture and Identity*. Srinagar: Centre of Central Asian Studies, University of Kashmir.

Harking, Tom. 2009. *Hard Truths about Uzbek Cotton, Los Angeles Times*, September 25, 2009, Opinion section. Available at http://articles.latimes.com/2009/sep/25/opinion/oe-harkin25

Held, David, Anthony McGrew, David Goldblatt, and Jonathan Perraton. 1999. *Global Transformation: Politics, Economics and Culture*. Cambridge: Polity Press.

Henry, Philip W. 2008. *Understanding Central Asia: Colonialism, Autonomy and Nationalism*. Rialto, CA: Author.

Herman, M. 1999. "Sustainable Agricultural Reform: The Case of Uzbekistan." In *Central Asia 2010: Prospects for Human Development*. pp. 84–95. New York: UNDP.

Hirst, Paul Q., and Grahame Thompson. 1996. *Globalization in Question: The International Economy and the Possibilities of Governance*. Cambridge: Polity Press.

Houseman, Ian. 2007. "Demonstration and Advisory Services Activities for Cotton Growing: A Case Study in Ak Altin, Uzbekistan." Proceedings of The Cotton Sector in Central Asia: Economic Policy and Development Challenges, conference held at SOAS, University of London, November 3–4, 2005 (coordinated by Deniz Kandiyoti).

Human Rights Watch. 2000. *Human Rights Watch World Report 2000: Uzbekistan*.

Huntington, Samuel. 1996. *The Clash of Civilizations and the Remaking of World Order*. New York: Simon and Schuster.

Hyman, Anthony. 1994. *Political Change in Post-Soviet Central Asia*. London: Russian and CIS Programme.

Hyman, Anthony. 1994. *Power and Politics in Central Asia's New Republics*. London: Research Institute for the Study of Conflict and Terrorism.

Ilkhamov, A. 1998. "Shikats, Dekhqon Farmers and Others: Farm Restructuring in Uzbekistan." *Central Asian Survey* 17, no. 4: 539–60.

Ilkhamov, A. 2007. "Use of Child Labour in the Cotton Sector of Uzbekistan: Agenda for Human Rights Advocacy." Presentation to the EJF Roundtable on Sustainable and Ethical Cotton, April 25.

ILO (International Labour Organization). 2010. *ILO Conventions on Child Labour*. Available at http://www.ilo.org/.

ILO. International Programme on the Elimination of Child Labour. 2002. *Combating Child Labour through Education*. Geneva. ILO.

International Cotton Advisory Committee (ICAC). 2010. *Cotton This Month: 4 January 2010*. Available at http://www.unctad.org/infocomm/anglais/cotton/Doc/ICAC0110.pdf.

International Crisis Group. 2005. *The Curse of Cotton: Central Asia's Destructive Monoculture*. Asia Report No. 93. Brussels: ICG.

International Crisis Group. 2009. *Central Asia: Migrants and the Economic Crisis*, Crisis Group Asia Report No. 183. Brussels: ICG.

International Crisis Group. 2005. *The Curse of Cotton: Central Asia's Destructive Monoculture*. Asia Report No. 93. Brussels: ICG.

International Labor Rights Forum and Human Rights Defenders in Uzbekistan. 2008. *Forced Child Labor in Uzbekistan's 2008 Spring Agricultural Season*. Available at http://www.laborrights.org/stop_child_labor/cotton_campaign/resources/1751.

International Labor Rights Forum. 2010. *Company Response to Forced Child Labor in Uzbek Cotton*. Available at http://www.laborrights.org/stop-child-forced-labor/cotton-campaign/company-response-to-forced-child-labor-in-uzbek-cotton.

International Labour Office. 2004. *Child Labour: A Text Book for University Students*. Geneva: ILO.

International Labour Rights Forum. 2008. *Forced Child Labour in the Cotton Industry of Uzbekistan*. Washington: Bama Athreya. Available at ipec/facts/ILOconventionsonchildlabour/lang--en/index.htm

Irin Asia. 2007. *Tajikistan: Poverty Fuels Child Labour*. May 1. Available at http://www.irinnews.org/report.aspx?ReportId=71890.

Kalb, Don. 2000. *Europeanization and Globalization*. Plovdiv Summer University.

Kandiyoti, D. 1996. "Modernization without the Market: The Case of the Soviet East." *Economy and Society* 25, no. 4: 529–42.

Kandiyoti, D. 1998. "Rural Livelihoods and Social Networks in Uzbekistan: Perspectives from Andijan." *Central Asian Survey* 17, no. 4: 561–78.

Kandiyoti, D. 2001. *Agrarian Reform, Gender and Land Rights in Uzbekistan*. New York: UNDP.

Kandiyoti, D. 2008. *Invisible to the World? The Dynamics of Forced Child Labour in the Cotton Sector of Uzbekistan*. London: Centre of Contemporary Central Asia and the Caucasus.

Karagiannis, Emmanuel. 2009. *Political Islam in Central Asia: The Challenge of Hizb ut-Tahrir*. London: Routledge.

Kavalski, Emilian (ed.). 2010. *Stable Outside, Fragile Inside? Post-Soviet Statehood in Central Asia*. Farnham, UK: Ashgate.

Kellner, Douglas. 1995. *Globalization and Postmodern Turn*. New York: Routledge.

Kemp, M. 1998. *Unveiling Uzbek Women: Liberation, Representation and Discourse, 1906–1929*. PhD Dissertation, University of Chicago, Chicago.

Khali, Tasneem. 2009. *Blood Cotton*. Independent World Report. Available at http://www.independentworldreport.com/2009/09/blood-cotton/.

Khan, A. R. 1996. The Transition to a Market Economy in Agriculture. In K. Griffin (ed.) *Social Policy and Economic Transformation in Uzbekistan*. Geneva: International Labour Office, and New York: United Nations Development Programme.

Khan, A. R., and D. Ghai. 1979. *Collective Agriculture and Rural Development in Soviet Central Asia*. London: Macmillan.

Kharimov, Islam A. 1998. *Uzbekistan on the Threshold of the Twenty-First Century: Challenges to Stability and Progress.* London: Palgrave Macmillan

Khojayori, Nasrullo, and Mikael Thompson. 2009. *Tajiki: Reference Grammar*. Washington, DC: Georgetown University Press.

Khojayori, Nasrullo. 2009. *Tajiki: An Elementary Textbook*, volumes 1 and 2. Washington, DC: Georgetown University Press.

Knowlton, Marylee. 2005. *Uzbekistan: Cultures of the World.* New York: Benchmark Books.

Kochhar, R. 1992. "International Labor Migration and Domestic Labor Supply." *Journal of Population Economics* 5.

Kucera, Joshua. 2009. "Eurasia Insight: Uzbekistan: US Legislators Press for Trade Penalties Against Tashkent." December 18, 2009, Eurasianet.org. Available at http://www.eurasianet.org/departments/insight/articles/eav121809a_pr.shtml 33.

Laruelle, Marlène. 2010. *Dynamiques migratoires et changements sociétaux en Asie centrale.* Paris: Petra Editions.

Lerman Z., and I. Stanchin. 2004. "Institutional Changes in Turkmenistan's Agriculture: Impacts on Productivity and Rural Incomes." *Eurasian Geography and Economics* 45, no. 1.

Lerman, Z. 1998. "Land Reform in Uzbekistan." In S. K. Wegren (ed.), *Land Reform in the Former Soviet Union and Eastern Europe*. London: Routledge.

Lerman, Z. 2008. *Tajikistan: An Overview of Land and Farm Structure Reforms*. Jerusalem: The Hebrew University of Jerusalem.

Levi, Scott C., and Ron Sela (eds.). 2009. *Islamic Central Asia: An Anthology of Historical Sources*. Bloomington: Indiana University Press.

Liebert, Saltanat. 2009. *Irregular Migration from the Former Soviet Union to the United States*. London: Routledge.

Lubin, N. 1984. *Labour and Nationality in Soviet Central Asia.* London: Macmillan.

Lubin. Nancy. 1994. *Leadership in Uzbekistan and Kazakhstan: Views of the Led*. Washington, DC: National Council for Soviet and East European Research.

Mackerras, Colin, and Michael Clarke (eds.). 2009. *China, Xinjiang and Central Asia: History, Transition and Crossborder Interaction into the 21st Century*. London: Routledge.

Macleod, Calcum, and Bradley Mayhew. 2002. *Uzbekistan: The Golden Road to Samarkand.* Hong Kong: Odyssey Publishers.

Malik, Hafeez. 1994. *Central Asia: Its Strategic Importance and Future Prospects.* New York: St. Martin's Press.

Manley, Rebecca. 2009. *To the Tashkent Station: Evacuation and Survival in the Soviet Union at War*. Ithaca, NY: Cornell University Press.

Mansoor, Ali, and Bryce Quillin. 2006. *Migration and Remittances: Eastern Europe and the Former Soviet Union*. Washington, DC: The World Bank.

Marat, Erica. 2010. *The Military and the State in Central Asia: from Red Army to Independence*. New York: Routledge.

Marketos, Thrassy N. 2008. *China's Energy Geopolitics: The Shanghai Cooperation Organization and Central Asia*. London: Routledge.

Mayhew, Bradley. 2007. *Central Asia: Lonely Planet.* London: Lonely Planet Publishers.

Mearns, R. September 1996. *Commons and Collectives: The Lack of Social Capital in Central Asia's Land Reforms.* Working Paper No. 40. Brighton, UK: Institute of Development Studies, University of Sussex.

Meldin, K. William, and William M. Cave. 1964. *Social Change and Education in Developing Areas: Uzbekistan, Comparative Education Review.* Chicago: University of Chicago Press.

Meldin, K. William, William M. Cave, and Finley Carpenter. 1971. *Education and Development in Central Asia: A Case Study of Social Change in Uzbekistan.* Leiden: E. J. Brill.

Melvin, Neil. 2003. *The Economy of Uzbekistan.* Regional Surveys of the World: Eastern Europe, Russia and Central Asia. 3rd. ed. London: Europa Publications.

Menon, Rajan. 1995. *The Dynamics of Security in Post-Soviet Central Asia.* Washington, DC: National Council for Research on Eastern Europe and Russia.

Mesbahi, Mohiaddin. 1994. *Central Asia and the Caucasus after the Soviet Union: Domestic and International Dynamics.* Gainesville: University Press of Florida.

Ministry of Agriculture of Kazakhstan. http://www.minagri.gov.kz/agro/?ID=5919.

Ministry of Foreign Affairs of the Republic of Uzbekistan. 2008. "Clarifications by the Ministry of Labor and Social Protection of Population of the Republic of Uzbekistan with Regard to "Forced Child Labor in Uzbekistan." April 9, 2008. Available at http://mfa.uz/eng/press.

Nazarov, Bakhtiyar A., and Denis Sinor. 1993. *Essays on Uzbek History, Culture, and Language.* Bloomington: Indiana University, Research Institute for Inner Asian Studies.

Neddermann, Hauke, 2009. *Sozialismus in Xinjiang: Das Produktions- und Aufbaukorps in den 1950er Jahren.* Berlin: Lit Verlag.

Nettleton, Susanna. 1992. "Uzbek Independence and Educational Change," *Central Asia Monitor* 3.

Nieuwenhuys, O. 1994. *Children's Lifeworlds: Gender, Welfare and Labour in the Developing World.* London: Routledge.

Nooshin, Luadan (ed.). 2009. *Music and the Play of Power in the Middle East, North Africa and Central Asia.* Farnham, UK: Ashgate.

Olcott, Martha B. 1995. *Central Asia's New States: Independence, Foreign Policy, and Regional Security.* Washington, DC: United States Institute for Peace Press.

Omelicheva, Mariya Y. 2010. *Counterterrorism Policies in Central Asia.* London: Routledge.

Osmonov, R. 2008. "Uzbekistan Signs a Cotton Contract Costing Nearly $1 Billion." *Turkish Weekly,* October 16. Available at http://www.turkishweekly.net/news/60257/uzbekistan-signs-a-cotton-contract-costing-nearly-1-billion.html.

Ostrowski, Wojciech. 2010. *Politics and Oil in Kazakhstan.* London: Routledge.

Ozcan, Gul Berna. 2010. *Building States and Markets in Central Asia: Enterprise Development in Central Asia.* Basingstoke, UK: Palgrave.

Paksoy, H. B., and Z. V. Togan. 1992. "The Origins of the Kazaks and the Ozbeks." *Central Asian Survey* 11, no. 3.

Patnaik, Ajay, and Tulsiram. 2010. *Globalisation and Eurasia.* New Delhi: KW Publishers.

Patzwaldt, K. 2004. *Labour Migration in Eastern Europe and Central Asia: Current Issues and Next Political Steps.* UNESCO Series of Country Reports on the Ratification of the UN Convention on Migrants. Paris: UNESCO.

Peyrouse, Sébastien. 2009. *The Multiple Paradoxes of the Agriculture Issue in Central Asia,* EUCAM Working Paper No. 6, available at http://papers.ssrn.com/sol3/papers.cfm?abstract_id=1603907

Poliakov, S. 1992. *Everyday Islam: Religion and Tradition in Rural Central Asia.* New York: M.E. Sharpe.

Pomfret R. 2007. "Using Energy Resources to Diversify the Economy: Agricultural Price and Distortion in Kazakhstan." *CASE Networks Studies and Analyzes* no. 335.

Pomfret, Richard W. T. 1995. *The Economies of Central Asia.* Princeton, NJ: Princeton University Press.

Prosser, Sarah. 2000. "Reform within and without the Law: Further Challenges for Central Asian NGOs." *Harvard Asia Quarterly* 4, no. 3: 4–16.

Racioppi, Linda, and Katherine O'Sullivan (eds.). 2009. *Gender Politics in Post-Communist Eurasia.* East Lansing: Michigan State University Press.

Rand, Robert. 2006. *Tamerlan's Children: Dispatches from Contemporary Uzbekistan.* Oxford: Oneworld Publications.

Rios, R. 2006. *Migration Perspectives: Eastern Europe and Central Asia.* Vienna: International Organization for Migration.

Rumar, Boris. 2003. *Central Asia in Transition: Dilemmas of Political and Economic Development.* Delhi: Aakar Books.

Rumer, B. Z. 1989. *Soviet Central Asia: A Tragic Experiment*. Boston: Unwin Hyman.
Rumer, Boris Z. 2000. *Central Asia and the New Global Economy: Critical Problems, Critical Choices*. Armonk, NY: M.E. Sharpe.
Rumer, Boris, and Stanislar Zhukov (eds.). 1998. *Central Asia: The Challenges of Independence*. London: Routledge.
Sahedeo, Jeff, and Russel Zanca. 2007. *Everyday Life in Central Asia: Past and Present*. Bloomington: Indiana University Press.
Saidazimova, G. 2008. "Uzbekistan: Cotton Industry Targeted by Child-Labor Activists." RFE/RL, January 19. Available at http://www.rferl.org/content/article/1079375. html.
Salimov, A. S., and A. Khaitov. 2004. *The Reorganization of Cooperative (Shirkat) Agricultural Institutions into Private Farms*. Tashkent: Center for Economic Research
Save the Children. 2002. *Child Labor in Uzbekistan*. Unpublished report. Tashkent: Save the Children.
Schillinger, W. F. 2003. *Cropping Systems Research Needs in Uzbekistan: A Report to the World Bank*. Washington, DC: World Bank.
Schoeberlein-Engel, John. 1996. "The Prospects for Uzbek National Identity." *Central Asia Monitor* 2.
Schroeder, G. 1996. "Economic Transformation in the Post-Soviet Republics." In B. Kaminski (ed.), *Economic Transition in Russia and the New States of Eurasia*. New York: M.E. Sharpe.
Sengupta, Anita, 2009. *Heartlands of Eurasia: The Geopolitics of Political Space*. Lanham, MD: Lexington.
Shagdar, Bolormaa. 2006. "Human Capital in Central Asia: Trends and Challenges in Education." *Central Asian Survey* 25, no. 4: 515–32.
Shimomura, Yasutami. 1994. "Transition to Market Economies in Asia." Economic Planning Agency. *The Economic Analysis* no. 137 (December).
Soros, George. 1998. *The Crisis of Global Capitalism: Open Society Endangered*. New York: BBS/Public Affairs.
Spoor, M. 2007. "Cotton in Central Asia: 'Curse' or 'Foundation for Development'?" In D. Kandiyoti (ed.), *The Cotton Sector in Central Asia: Economic Policy and Development Challenges*. London: School of Oriental and African Studies.
Spoor, Max. 1993. *Transition to Market Economies in Former Soviet Central Asia: Dependency, Cotton, and Water*. The Hague: Institute of Social Studies.
Spoor, Max. 1995. "Agrarian Transition in Former Soviet Central Asia: A Comparative Study of Uzbekistan and Kyrgyzstan." *Journal of Peasant Studies* 23, no. 1: 46–63.
Standing, G. 1996. *Russian Unemployment and Enterprise Restructuring: Reviving Dead Souls*. London: Macmillan.
Starr, S. Frederick. 2005. "A Partnership for Central Asia." *Foreign Affairs* 84 (July/August), 4.
Stern, D. L. 2008, "Tajik Farmers Enslaved Where Cotton Is." *Farmblogs*, October 15. http://farmblogs.blogspot.com/2008/10/tajik-farmers-enslaved-wherecotton-is.html.
Stern, David L. 2008. "In Tajikistan, Debt-Ridden Farmers Say They Are the Pawns." *New York Times*, October 14, 2008. Available at http://www.nytimes.com/2008/10/15/world/asia/15tajik.html?_r=1.
Strolz, Herwig M. 2005. "Perspectives for Textile Industrialization in the Post-Quota Era." Proceedings of the International Cotton Conference Cotton and Textiles—Important Factors of Economic Growth in Central Asia, Tashkent, Uzbekistan, May 24–25, 2005.
Technical Assistance to the Commonwealth of Independent States (TACIS) and Government of Uzbekistan. 1996. *Pilot Integrated Development Programme, Bulungur District, Samarkand, Uzbekistan*. Tashkent: Final Report BS6-Land Tenure.
Thurman, J. M. 1999. *The Command-Administrative System in Cotton Farming in Uzbekistan 1920s to Present*. Papers on Inner Asia No. 32. Bloomington: Indiana University.
Tismaneanu, Vladmir. 1995. *Political Culture and Civil Society in Russia and the New States of Eurasia*. Armonk, NY: M.E. Sharpe.

Tokhtahodjaeva, M. 1995. *Between the Slogans of Communism and the Laws of Islam*. Lahore: Shirkat Gah.

Tranum, San. 2010. *Daily Life in Turkmentbashy's Golden Age: A Methodologically Unsound Study of Interactions between the Tribal Peoples of America and Turkmenistan*. Privately published, USA.

Trevisani, T. 2007. "The Emerging Actor of Decollectivization in Uzbekistan: Private Farming between Newly Defined Political Constraints and Opportunities." In D. Kandiyoti (ed.), *The Cotton Sector in Central Asia: Economic Policy and Development Challenges*. London: School of Oriental and African Studies

Trushin, E. 1998. "Uzbekistan: Problems of Development and Reform in the Agrarian Sector." In B. Rumer and S. Zhukov (eds.), *Central Asia: The Challenges of Independence*. New York: M.E. Sharpe.

*Turkish Weekly*. 2004. "Seven Year Olds Pick Uzbek Cotton." November 22. Available at http://www.turkishweekly.net/news/412/seven-year-olds-pick-uzbek-cotton.html.

U.N. Development Project. 1997. *Human Development Report: Uzbekistan*. New York: UNDP.

Umarov Kh., and M. Gafurov, 2008. "Finansovyi rynok i khlopovaia problema." *East Time*, November 11. Available at http://www.easttime.ru/analitic/1/10/524.html.

UNICEF (with ILO and the Government of Uzbekistan). 2005. *Rapid Assessment of Informal Employment of Children in Urban and Rural Areas of Uzbekistan*, Tashkent: UNICEF.

UNICEF. 2008. *Multiple Indicator Cluster Surveys (MICS): Questionnaire*. Available at http://www.childinfo.org/mics3_questionnaire.html.

UNICEF. 2010. *Child Rights: The Best Possible Start in Life*. Available at http://www.unicef.org/uzbekistan/rights.html.

United Nations Development Programme. 2001. *Millennium Development Goals*. New York: UNDP.

Uzbekistan National News Agency. 2008. "Measures on Implementation of Convention on Minimum Age for Acceptation for Employment and Convention on Prohibition and Immediate Actions for the Eradication of the Worst Forms of Child Labor Ratified by the Republic of Uzbekistan." Available at http://uza.uz/en/documents/412/

van den Broeck, J. 1996. *The Economics of Labour Migration*. Brookfield, VT: Edward Elgar.

Vassiliev, Alexei. 2001. *Central Asia: Political and Economic Challenges in the Post-Soviet Era*. London: Saqi Books.

Ware, Robert Bruce, and Enver Kisriev. 2010. *Dagestan: Russian Hegemony and Islamic Resistance in the North Caucasus*. New York: M.E. Sharpe.

Waters, Malcolm. 1995. *Globalization*. London: Routledge.

Watters, K. 1990. "The Current Family Planning Debate in Soviet Central Asia." *Central Asian Survey* 9, no.1: 5–86.

Weiner, M. 1991. *The Child and the State in India*. Princeton, NJ: Princeton University Press.

Wooden, Amanda E., and Christoph H. Stefes (eds.). 2009. *The Politics of Transition in Central Asia and the Caucasus: Enduring Legacies and Emerging Challenges*. London: Routledge.

World Bank. 2006. *Asian Development Indicators 2005*. Asian Development Outlook 2006. Washington, DC: World Bank.

World Bank. 1993. *Uzbekistan: Plan for Economic Reform*. Washington, DC: World Bank.

World Bank. 1999. *Uzbekistan: Social and Structural Policy Review*. Report No. 19626. Washington, DC: World Bank.

World Bank. 2008. *Uzbekistan at a Glance*. Available at http://devdata.worldbank.org/AAG/uzb_aag.pdf

World Bank. 2010. *Uzbekistan: Country Brief 2010*, Available at http://go.worldbank.org/3L4P0K4NQ0

Yemelianova, Galina M. (ed.). 2010. *Radical Islam in the Former Soviet Union*. London: Routledge.

Zaeef, Abdul Salam, 2010. *My Life with the Taliban* (translated from the Pashto and edited and introduced by Alex Strick van Linschoten and Felix Kuehn; foreword by Barnett R. Rubin). New York: Columbia University Press/London: Hurst and Co.

Zhores, Madvedev. 1996. *" Soviet Agriculture": Issues in Development Discussion Paper 14*. Development and Technical Cooperation Department. Geneva: International Labour Office.

Zviagelskaia, I. D. 1995. *The Russian Policy Debate on Central Asia*. London: Royal Institute of International Affairs.

# Index

# About the Author

**Bilal Bhat** holds a doctorate degree in sociology from the Centre of Central Asian Studies, University of Kashmir, J&K, India. He was awarded Post-Doctorate Research Fellowships by the University Grants Commission of India and the Indian Council of Social Science Research, Delhi. In addition to his books *Rehabilitation of Child Labour: Problems and Prospects* and *Aspirations and Challenges of Youth*, Bilal Bhat has contributed a number of research articles on South and Central Asia, particularly on Uzbekistan in various reputable international journals. He has also carried out extensive field work in various regions of Central Asia especially Tashkent, Samarkand, Bukhara, Urgench, Khiva, Namangan, Andijan, and the Fergana Valley of Uzbekistan on various themes. He presently works as assistant professor of research.